OCCASIONAL PAPER

RAND COUNTERINSURGENCY STUDY • PAPER 3

Understanding Proto-Insurgencies

Daniel Byman

Prepared for the Office of the Secretary of Defense

Approved for public release; distribution unlimited

NATIONAL DEFENSE RESEARCH INSTITUTE

The research described in this report was prepared for the Office of the Secretary of Defense (OSD). The research was conducted in the RAND National Defense Research Institute, a federally funded research and development center sponsored by the OSD, the Joint Staff, the Unified Combatant Commands, the Department of the Navy, the Marine Corps, the defense agencies, and the defense Intelligence Community under Contract W74V8H-06-C-0002.

Library of Congress Cataloging-in-Publication Data

Byman, Daniel, 1967-
Understanding proto-insurgencies / Daniel L. Byman.
p. cm. — (Rand counterinsurgency study ; paper 3)
Includes bibliographical references.
ISBN 978-0-8330-4136-4 (pbk. : alk. paper)
1. Counterinsurgency. 2. Insurgency. 3. Terrorism—Prevention. 4. United States—Foreign relations.
5. United States—Influence. I. Title.

U241.B95 2007
355.02'18—dc22

2007034546

The RAND Corporation is a nonprofit research organization providing objective analysis and effective solutions that address the challenges facing the public and private sectors around the world. RAND's publications do not necessarily reflect the opinions of its research clients and sponsors.

Published 2007 by the RAND Corporation
1776 Main Street, P.O. Box 2138, Santa Monica, CA 90407-2138
1200 South Hayes Street, Arlington, VA 22202-5050
4570 Fifth Avenue, Suite 600, Pittsburgh, PA 15213-2665
RAND URL: http://www.rand.org/
To order RAND documents or to obtain additional information, contact
Distribution Services: Telephone: (310) 451-7002;
Fax: (310) 451-6915; Email: order@rand.org

Preface

The study reported here was undertaken as part of a RAND Corporation research project for the U.S. Defense Department on how to improve U.S. counterinsurgency (COIN) capabilities. It should be of interest to the U.S. government and to other countries and organizations now rethinking COIN strategies and retooling COIN capabilities in view of developments since September 11, 2001, as well as to scholars trying to understand continuity and change in this field.

The larger RAND project will yield a stream of interim products during its course. It will culminate in a final report that draws on that stream of work. Thus, this report can and should be read both as an output, in and of itself, and as a piece of an emerging larger picture of COIN.

The process by which small terrorist groups and insurrections transition to full-blown insurgencies has received only limited scrutiny. However, these groups are highly vulnerable in their early stages, and it would save many lives and be far more cost-effective to fight them before their ranks swell. This paper is intended to both focus attention on this early stage of insurgency and help identify capabilities for combating potential insurgencies before they become full-blown.

This analysis was conducted within the International Security and Defense Policy (ISDP) Center of the RAND National Defense Research Institute, a federally funded research and development center sponsored by the Office of the Secretary of Defense, the Joint Staff, the unified combatant commands, the Department of the Navy, the Marine Corps, the defense agencies, and the defense intelligence community.

For more information on RAND's ISDP Center, contact the director, James Dobbins. He can be reached by email at james_dobbins@rand.org; by phone at 703-413-1100, extension 5134; or by mail at RAND Corporation, 1200 South Hayes Street, Arlington, VA 22202-5050. More information about RAND is available at www.rand.org.

Contents

Summary

Small bands of fighters and terrorist groups usually seek to become full-blown insurgencies as part of their strategy for victory. But their task is difficult. The groups often start out with few members, little funding, and limited recognition, while the governments they oppose enjoy coercive and financial advantages and are seen as legitimate by most domestic and international audiences. Despite these difficulties, some groups do make the successful transition to full-blown insurgency. That transition is the focus of this paper.

The Tasks Before the Proto-Insurgent

To gain the size and capabilities of an insurgency, a terrorist group or other would-be insurgent movement must take several steps:

- First, proto-insurgents must create a politically relevant identity—a surprisingly difficult task. Success requires undermining rival identities put forward by the state or other groups and convincing people that the group comprises Muslims, Kurds, or whatever particular identity it champions. This identity is the basis for a group's subsequent organization and expansion.
- Second, the identity must be linked to a cause that is popular beyond the terrorist group or band of insurrectionists. Many causes championed by proto-insurgents have little inherent popularity, and governments can often co-opt the more popular elements of a cause. Nationalism is perhaps the most potent cause to harness.
- Third, the proto-insurgents must gain dominance over their rivals. The primary foe at this early stage is not the government, but the welter of rival organizations that compete for recruits and money. Many of these organizations seek to exploit the same cause as does the proto-insurgent, but they use a different identity or platform to do so. Not surprisingly, proto-insurgent energies are often consumed by fights within their own community.
- Finally, proto-insurgents need a respite from police, intelligence, and military services. Many groups thus find that a sanctuary or "no go" zone is often essential for their survival.

Violence is instrumental in all the tasks proto-insurgencies seek to accomplish. Violence can aid recruitment, attracting media attention and separating the group from more peaceful, and thus more accommodating, rivals. Even when it fails to inspire, violence can intimidate citizens into supporting the would-be insurgents, or at least not supporting the government. Violence also forms a bond within the group and makes the moderate option less tenable. In part, this occurs through intimidation: Moderates are often the first target for radical groups. Also, the climate of violence over time makes promises of moderation wear thin. Perhaps most important, violence reduces confidence in government administrative and police structures by demonstrating that the government cannot fulfill its most essential task, that of protecting the citizenry.

Violence, however, can also backfire on the proto-insurgents. Simply put, few people support violence. Thus proto-insurgents face the dilemma of deciding whether and how much to target civilian populations. They seek to provoke a reaction from a government or from rival groups but not to alienate their constituencies.

The success or failure of a proto-insurgent movement depends only in part (and at times only in small part) on its own campaign. The reaction of the state is often the most important factor in a movement's overall success. In particular, states can disrupt organizations through various forms of policing and repression and can co-opt potential leaders and make them allies of the state. States can also divide the identity that proto-insurgents wish to put forward.

Some regimes are unable to implement effective repressive measures, while others can capitalize on circumstances that increase the effectiveness of control. The resources of the would-be insurgent group, the level of outside backing, its access to arms, its social structure, and the type of regime are among the many factors that affect the effectiveness of control. Governments with a high degree of popular support and skilled police and intelligence services are more likely to implement control policies more effectively.

The Role of Outside Support

State support offers numerous advantages to groups seeking to become insurgencies. Outside states can offer a safe haven that groups often lack. They can also offer money, training, and help with political mobilization. State support also helps groups overcome logistical difficulties and hinders intelligence-gathering against them. Finally, state support legitimizes a proto-insurgent group and makes government delegitimization efforts almost impossible. Such backing can help a group resist government counterterrorism and counterinsurgency (COIN) campaigns.

Despite all these benefits, state support is a mixed blessing. Even the most supportive and ideologically sympathetic regimes have their own distinct national interests and domestic politics, making them unlikely to completely embrace the proto-insurgents' agenda. Proto-insurgents may also lose freedom of action, as states fear risking an all-out confrontation and thus put a brake on their proxies. At times, a purported backer will deliberately try to weaken the overall movement, even as it supports particular groups. Finally, outside support can hinder a group's effort to harness nationalism.

Defeating Proto-Insurgencies

Governments opposing proto-insurgencies must recognize the proto-insurgents' many weaknesses and avoid an overreaction that may inadvertently strengthen the group. Perhaps the best and most efficient way to prevent proto-insurgents from gaining ground is through "in-group" policing. Individual communities know their own members, particularly in tight-knit societies in the developing world. These communities thus make intelligence-gathering easier and enable the use of arrests or other forms of pressure with far greater discrimination. In-group policing, of course, requires a government to work with more-moderate members of a community and often to make political concessions to them.

States can also promote rival identities. Ideally, states can build up the national identity to increase bonds among citizens. They can use powerful tools such as education, control of language, and influence over the media to build a rival identity to the one the proto-insurgents support. Alternatively, they can also divide the identity of the proto-insurgent movement.

The most obvious action for the United States to take in its COIN campaign is to anticipate the possibility of an insurgency developing before it materializes. Many of the recommended steps are relatively low cost and easy to implement, especially when compared with fighting a full-blown insurgency. Indicators of potential insurgencies are discussed in Appendix C.

It is often important for the United States to stay in the background when dealing with potential insurgencies. Since the best cause for insurgents to harness is nationalism, direct and open U.S. support of a government can undercut that government's legitimacy. The United States can, however, provide training and advisory programs that are largely behind the scenes, particularly if they are conducted outside the country. These programs should focus on improving indigenous capabilities rather than on substituting U.S. roles for them.

Building a strong police force is also important—usually much more important than aiding the military. Police typically are far better suited to defeating small groups, because they know the communities well and are trained to use force discriminately. Success in defeating insurgent movements requires not only that the police be strong and numerous, but that the laws they enforce be suited for counterinsurgency. Thus, legal reform is a vital early step in counterinsurgency. Creating programs that combine the talents of military officers, police officials, and intelligence professionals would be an important step forward in fighting proto-insurgents.

Intelligence should also be redirected to focus on the conditions that foster insurgency as well as on the presence of insurgencies. Once an insurgency is manifest, it is far harder to defeat. Identifying would-be insurgents and assessing their chances of success thus is a highly effective way to help direct resources.

Finally, the United States can help inhibit outside support for proto-insurgencies. U.S. power is often better directed at coercing hostile states than it is at directly fighting insurgents within friendly countries.

Acknowledgments

This paper benefited considerably from the insights of many people at RAND and elsewhere. RAND colleagues Peter Chalk, David Gompert, John Gordon, Seth Jones, Martin Libicki, William Rosenau, and Lesley Warner all added valuable insights, as did several U.S. government officials. Particular thanks are due to Brian Jenkins and Jeremy Shapiro, both of whom went above and beyond the usual reviewer duties, providing unstinting criticisms and detailed suggestions that greatly improved the overall quality of the work.

Abbreviations

Amal	Afwaj al-Muqwama al-Lubnaniya
CIA	Central Intelligence Agency
COIN	counterinsurgency
EIJ	Egyptian Islamic Jihad
ETA	Euskadi Ta Askatasuna
FARC	Revolutionary Armed Forces of Colombia
IDF	Israel Defense Forces
IG	Islamic Group
IMU	Islamic Movement of Uzbekistan
IRGC	Islamic Revolutionary Guard Corps
JKLF	Jammu and Kashmir Liberation Front
LTTE	Liberation Tigers of Tamil Eelam
MEK	Mujahedin-e Khalq
PIRA	Provisional Irish Republican Army
PKK	Kurdistan Workers Party
PLO	Palestine Liberation Organization
QAP	Al-Qa'ida of the Arabian Peninsula

CHAPTER ONE

Introduction

Pity the would-be insurgent. He and his comrades are unknown to the population at large, and their true agenda has little popularity. Indeed, most countries around the world oppose their agenda. Many of the fighters are not experienced in warfare or clandestine operations, making them easy prey for the police and intelligence services. Their families are at the mercy of government security forces. The government they oppose, in contrast, is relatively rich, has thousands or even millions of administrators, policemen, and soldiers, and enjoys considerable legitimacy. As J. Bowyer Bell argues, "The assets of the state are so apparent, so compelling, so easy to number, chart, and grasp, and the prospects of the rebel are so faint, that only the most optimistic risk an armed struggle" (Bell, 1994, p. 115).

Despite this barren soil, some insurgencies blossom. Indeed, many of the most important "terrorist" groups in the world—including the Lebanese Hizballah, the Liberation Tigers of Tamil Eelam (LTTE), and the Revolutionary Armed Forces of Colombia (FARC)—are better described as insurgencies that use terrorism than as typical terrorist movements. But this blossoming did not happen overnight. These groups began as small bands of terrorists who, over time, took on the trappings of guerrilla armies. Still other groups try to avoid relying exclusively on terrorism at any stage in their development, employing small-group guerrilla tactics in the hopes of sparking a broader rebellion.

The problem of these developing insurgencies is central to the U.S. war on terrorism today. In locales ranging from southern Thailand to Saudi Arabia to Morocco, self-styled "armies" exist that are in reality small terrorist groups or weak guerrilla movements seeking to create a broader social and guerrilla movement. Their goal is to create an insurgency where none exists, using terrorism as a weapon.

The flowering of insurgencies does not occur instantly or automatically. Would-be insurgents often fail repeatedly or succeed only partly. For every group that becomes an insurgency, dozens—or perhaps hundreds—fail. Bell notes that the Cuban revolution inspired more than 200 insurrections, and they all failed (Bell, 1994, p. 115). Most of the groups Bell refers to were rural guerrilla groups, but this fate is also common to terrorist groups. Terrorism expert Bruce Hoffman notes that approximately 90 percent of all terrorist groups collapse within a year, and only half of the hardy remainder make it through another decade (Hoffman, 2002a, p. 84). Even a cursory look at the trajectory of successes such as the LTTE or Hizballah indicates that they could easily have collapsed or had rivals supersede them.

This paper examines the question of how small groups, whether terrorists or very weak guerrilla movements, become larger insurgencies. It looks at the factors that contribute to the success of a movement, which include ideational politics, using violence without alienating potential constituents, managing a state sponsor, and government response. These factors are described analytically and with examples, but it is important to note that there is no recipe for success. Most of the factors involve both benefits and risks.

The consequences of failing to consider the many conditions that contribute to the development of insurgencies can be devastating. Governments have at times overreacted to terrorist groups or small bands of guerrillas, not recognizing that the conditions for them to ignite a broader conflagration are lacking—unless the government's heavy hand creates popular anger where none previously existed. Complacency can be equally dangerous. At times, governments have failed to crush small groups when they had the chance. Hoffman laments that after the invasion of Iraq in 2003, the United States failed to detect the formation of an insurgency until it was too late, making victory far more difficult or perhaps even impossible (Hoffman, 2004, pp. 2–4).[1]

The remainder of this paper is organized as follows. Chapter Two looks at the nexus between terrorism and insurgency, noting where they overlap and where they differ and how proto-insurgencies fit into the picture. Chapter Three focuses on the role of violence in the process of insurgency development. Chapter Four examines the basic needs of insurgent movements, most or all of which are initially lacking. With this background, Chapter Five explores how small groups, usually terrorist groups but sometimes small bands of guerrillas, try to create an insurgency and how states often mishandle proto-insurgents and make the problem worse. Chapter Six concludes by looking at the means of fighting proto-insurgencies, which differ considerably from standard counterinsurgency (COIN) operations. Appendix A looks at the experiences of three armed groups—the Lebanese Hizballah, Egyptian Islamic Jihad (EIJ) and the Islamic Group (IG), and the Palestinian organization Fatah until 1993—and examines why Hizballah became a successful insurgent movement, while EIJ and Fatah never made this leap despite being enduring terrorist groups. Appendix B attempts to apply the ideas in this paper prospectively, examining the possibility of an insurgency in Saudi Arabia today. Finally, Appendix C describes some measures that indicate when proto-insurgencies may grow into full-blown insurgent movements.

---

[1] This failure was particularly pronounced on the policy side, but much of the senior military leadership was blind to the possibility of, and then the emergence of, a full-blown insurgency (see Ricks, 2006).

CHAPTER TWO

Terrorism, Insurgency, and Proto-Insurgency

The overlap between insurgency and terrorism has important implications for both effective counterterrorism and effective COIN operations.[1] Proto-insurgencies are often found at this nexus.

Not all terrorist groups are insurgencies, but almost every insurgent group uses terrorism. Although the exact percentage depends heavily on coding decisions, in my judgment approximately half of the groups listed by the U.S. Department of State as Foreign Terrorist Organizations are insurgencies as well as terrorist groups. Even more important, the majority of the most worrisome terrorist groups in the world today are also insurgencies. The LTTE, the Kurdistan Workers' Party, the Lebanese Hizballah, and the FARC all use guerrilla war as a major component in their struggles. Moreover, several leading analysts consider al-Qa'ida to also be essentially an insurgency (Anonymous, 2003, p. xviii).[2] Indeed, many terrorist groups that did not use guerrilla warfare, including the Provisional Irish Republican Army (PIRA) and Hamas, attempted to do so but found they were not strong enough.

This report uses the following definition of insurgency, provided in the Central Intelligence Agency (CIA) pamphlet *Guide to the Analysis of Insurgency*:[3]

[1] This chapter draws heavily on *Deadly Connections: States That Sponsor Terrorism* (Byman, 2006). Later sections also draw in part on this book, particularly the section on the impact of external support. Parts of the section in Chapter Five on when control works (p. 21) come from *Keeping the Peace: Lasting Solutions to Ethnic Conflict* (Byman, 2002). The Hizballah section in Appendix A draws in part on a forthcoming book chapter prepared for the United States Institute of Peace.

[2] The author of this source, revealed after its publication to be Michael Scheuer, contends that bin Laden is promoting (and at times directing) a "worldwide, religiously inspired, and professionally guided Islamist insurgency." Much of al-Qa'ida's activities are also designed to establish new or bolster existing insurgencies by providing them with money, supplies, inspiration, and training. Both the Provisional Irish Republican Party (PIRA) and Hamas have elements of an insurgency, though neither "controls" territory in a manner comparable to that of the Lebanese Hizballah or the FARC. In my judgment, bin Laden's objectives often coincide with those of many insurgent leaders. He seeks to control territory and capture states. In addition, he is keen to use the actions of the governments he opposes against them, a classic insurgent technique. Although the United States understandably focuses on al-Qa'ida's terrorist activities, much of the organization's money and energy has historically gone into fighting in local insurgencies and proselytizing. However, despite this insurgent-like emphasis, al-Qa'ida does many things that are atypical of an insurgent movement. In particular, it seeks to foster groups and individuals around the world that are not affiliated directly with a particular movement but that wish to carry out individual activities as jihad.

[3] The pamphlet was published in the 1980s. This definition is more comprehensive than others, but they too emphasize the importance of guerrilla warfare. Fearon and Laitin see insurgency as involving "small, lightly armed bands practicing guerrilla warfare from rural base areas" (Fearon and Laitin, 2003).

> Insurgency is a protracted political-military activity directed toward completely or partially controlling the resources of a country through the use of irregular military forces and illegal political organizations. Insurgent activity—including guerrilla warfare, terrorism, and political mobilization, for example, propaganda, recruitment, front and covert party organization, and international activity—is designed to weaken government control and legitimacy while increasing insurgent control and legitimacy. The common denominator of most insurgent groups is their desire to control a particular area. This objective differentiates insurgent groups from purely terrorist organizations, whose objectives do not include the creation of an alternative government capable of controlling a given area or country (CIA, n.d., p. 2).

By this definition, insurgencies typically, though not inherently, have three components: political mobilization, guerrilla warfare, and the use of terrorism. An insurgent group may use terrorism, but it does not necessarily do so. It is analytically possible (though empirically rare) for an insurgent group to use only guerrilla warfare and political mobilization, and not terrorism. A guerrilla group could focus on military targets and others involved in a COIN campaign. Noncombatants might be killed, but the group's actions would not be terrorism if they were a by-product of a military campaign and thus not intended to send a broader political message (a characteristic part of most definitions of terrorism).[4]

Groups' organizational structures and preferred methods tend to reflect whether guerrilla war or terrorism is their primary technique. For example, groups organized into irregular military units are more likely to pursue guerrilla war, while those with smaller cell structures probably intend to use terrorism. However, some organizations incorporate both structures. The Lebanese Hizballah, for example, has distinct components for waging guerrilla war and for conducting terrorist attacks.

It is particularly important to recognize that atrocities that are part of a guerrilla struggle are not necessarily terrorism. Almost all guerrilla armed forces commit some atrocities against civilians, such as rape, murder, and plundering. These atrocities may have political ramifications, but if their purpose is not political or intended to influence a broader audience, they should not be considered terrorism. However, the same acts, if they are intended to send a political message (such as discouraging collaboration or prompting ethnic cleansing), would be terrorism as well as part of a guerrilla war.

Size is also part of the insurgent picture. Although they can use terrorism, small groups cannot effectively wage guerrilla war and conduct widespread political mobilization. The Red Army Faction in Germany found it impossible to use guerrilla warfare and difficult to mobilize people on a mass scale, in contrast to Pakistan's Lashkar-e Tayyiba, which has thousands of members and part-time supporters. However, larger groups can also have military challenges. Some groups have admitted too many members too quickly, resulting in a force that was large but poorly trained and thus ineffective. Nevertheless, other things being equal, larger size is a great benefit in insurgent war.

---

[4] Hoffman notes that terrorism is "designed to have far-reaching psychological repercussions beyond the immediate victim or target" (Hoffman, 2006, p. 40).

In order to gain size, proto-insurgencies focus on steps that can help them become full-blown insurgent movements, such as creating and strengthening the political identity and cause they champion, eliminating competitors, and finding a sanctuary. Terrorism can help with these steps, but it is not necessary. The size at which a proto-insurgency is better classified as a full-blown insurgency is best seen relative to the size of the state's population and the strength of its government. A force of 500 fighters would be quite large against a weak government like that of Tajikistan or in a tiny country such as Brunei, but it would be rather small in a country as large as China.

With size comes another common characteristic of an insurgency: the ability to seize and hold territory. Not only is this ability a key marker for the overall success of a movement, it also has important organizational benefits. As noted below, control of territory conveys huge rewards with regard to recruitment and avoiding a government's counterterrorism campaign.

Thus, it is important to distinguish terrorist groups that are also insurgencies from those that are not.[5] Some groups, such as the Burundian Hutu marauders, are primarily guerrilla groups, focusing their effort on enemy government forces. Others, such as the LTTE and various Kashmiri militants, use guerrilla tactics and terrorism simultaneously. Still others, such as Hamas, rely primarily on terrorism to advance their cause. A few, including the Greek November 17 Organization, rely entirely on terrorism and do not engage in political mobilization or guerrilla war of any sort. Toward the "terrorism" end of the scale, group size shrinks, and little, if any, territory is controlled. However, it is important to recognize that this distinction is not a dichotomy: Using my definitions, it is possible to have a "pure" terrorist group or a "pure" insurgency, but in many instances—many of which involve the most dangerous terrorist groups—the actors involved are insurgent groups that regularly use terrorism as a tactic.

The proto-insurgency, then, is a small, violent group that seeks to gain the size necessary to more effectively achieve its goals and use tools such as political mobilization and guerrilla warfare as well as terrorism. The group can already use terrorism to this end and can conduct political mobilization, but only on a small scale. It does not have to use terrorism to be a proto-insurgency—a small guerrilla group operating in a limited area would fall into this category as well. In either case, the group sees becoming an insurgency as important to its strategy: It will "win" by mobilizing the people and conducting guerrilla war, either to defeat the state outright or to force its collapse through protracted warfare. Thus, a group like Japan's Aum Shinrikyo, despite its size, would not be an insurgency or proto-insurgency because guerrilla warfare was not part of its strategy. Conceptually, a proto-insurgency is what exists before Mao's "Phase I": the strategic defensive. The "party" that Mao and others sought to expand is largely nonexistent for proto-insurgencies, and they cannot spare the cadre to infiltrate other social movements.[6]

[5] The U.S. government definition of terrorism, which includes military forces not engaged in combat as "noncombatants" and also defines intelligence and law enforcement personnel as noncombatants, effectively excludes any possibility of distinguishing between the two. Any inhibitions that insurgent groups might have are further reduced by definitions that lump almost all guerrilla activity under the rubric of terrorism. A group that attacked only soldiers would still be depicted as a terrorist group.

[6] See Beckett (2003, pp. 73–75) for a discussion of Mao's views.

The proto-insurgency is a key stage.[7] The vast majority of terrorist groups are defeated within their first few years of existence; only a few survive to pose a serious threat, and of those, only a few become insurgencies (Hoffman, 2002a, p. 84). As this suggests, proto-insurgencies have considerable vulnerabilities that can be exploited.

Members of a proto-insurgency, like most terrorists, tend to be characterized by conviction and idealism—traits not always shared as the organization gets larger. As Hoffman notes, "The terrorist is fundamentally an *altruist:* he believes that he is serving a 'good' cause designed to achieve a greater good for a wider constituency" (Hoffman, 2006, p. 43). Insurgencies must look for both leaders and followers, often in different places. Many leaders come from what David Galula styles the "rejected elite" (Galula, 1964, p. 22). They are better educated than most of the population, and they have aspirations to leadership that are often unfulfilled through the current system. Followers can come from many sources and are wooed not only through ideology, but also through promises of future rewards or coercion against them and their families.

---

[7] In its *Guide to the Analysis of Insurgency*, the Central Intelligence Agency notes that this is a common stage for insurgency (CIA, n.d., p. 3).

CHAPTER THREE

The Role of Violence

By definition, terrorism, guerrilla warfare, and insurgencies are steeped in violence. Violence is instrumental for proto-insurgencies in all the tasks they seek to accomplish: It can foster an identity, create a cause, outpace rivals, attract outside support, and—most important—lead a state to overreact. Thus it is not surprising that terrorists typically see violence as a tool, perhaps *the* tool, for the creation of insurgencies. As Carlos Marighela, the Brazilian leader who authored the *Minimanual of the Urban Guerrilla*, wrote, "Action creates the vanguard."[1]

Violence can aid recruitment. For Marighela and others, violence is a form of propaganda. Such "propaganda by the deed" is intended to both educate the uncommitted on the cause in general and inspire them to act (Hoffman, 2006, p. 17). Violence attracts media attention, and thus the group serves as a magnet for like-minded fighters.

Once fighters are recruited, violence also forms a bond within the group. Because it is illegal and commonly perceived as immoral, using violence separates group members from others who, in group eyes, only talk and do not act. In addition, the use of violence has a "no going back" quality—group members who could have defected and lived in peace now are bound to the group, in part because they cannot return to civilian life.

Violence reduces confidence in government administrative and police structures by demonstrating that the government cannot fulfill its most essential task, i.e., protecting the citizenry. Violence thus demonstrates that the terrorists or small guerrilla bands are not fated to lose, perhaps the most daunting perception they must overcome. Thus, they may find that resources are more likely to be forthcoming and that individuals will fear denouncing them because of possible later consequences (Leites and Wolf, 1970, p. 13).

Violence also sets a group apart from its rivals. At times, dozens of cells are competing for recruits and money, and a group that can successfully pull off attacks in the face of state opposition demonstrates its prowess to would-be recruits. Violence committed in order to gain support must be calculated to win the admiration of extremists, something that requires restraint as well as daring. In addition, violence must capture media attention (or, increasingly, be exploitable by the proto-insurgents' own media capabilities). Only then will the recruitment dimension of a violent act have its full effect.[2]

[1] As quoted in Crenshaw (1985, p. 475).

[2] See Hoffman (2006, pp. 173–228) for the role of the media and how new technologies have changed it.

Violence makes the moderate option less tenable. In part, this is because of intimidation: Moderates are often the first target for radical groups. Also, the climate of violence over time makes promises of moderation wear thin. In Iraq today, for example, slogans in Shi'a areas proclaim that "Sistani is sleeping"—in essence, denouncing the revered Shi'a leader for his perceived passivity in the face of anti-Shi'a violence (International Crisis Group, 2006b, p. 25). In general, creating a democratic system is exceptionally difficult when violence is rampant, as the trust needed for political leaders to come together is lacking.

Violence not only inspires, it also intimidates. Citizens who might sympathize with the would-be insurgents' cause may inform on them because they believe that the state will punish them for collaborating if they do not. Early on, rebels must obtain a degree of insulation from a public that may denounce them, whether through sympathy, remoteness, or intimidation. Leites and Wolf refer to this problem as "nondenunciation"—convincing the population not to inform on rebel activities to the government (Leites and Wolf, 1970, p. 10).

Perhaps most important, violence can undermine the ability of the state to rule and can gain the group tactical advantages in the broader political-military struggle. For example, attacks on civilians may lead a rival ethnic group to flee a contested area. Strikes on government officials may make an area ungovernable, demonstrating that the government cannot protect its people and provide for their welfare while convincing other officials to collaborate.[3]

Proto-insurgents must decide not only which local targets to hit, but also whether to focus on local targets or conduct international terrorism. Going international is risky, but at times it has rewards. Attacking U.S. or other targets outside the immediate theater of operations can lead to increased U.S. support for the government and to a denial of international aid to the would-be insurgents. The United States provided considerable direct and indirect support to the Philippine government because the Abu Sayyaf Group had conducted attacks on Westerners and was linked to bin Laden—attacks and connections that, for the group, cost it far more than it gained relative to Manila.

However, international terrorism generates tremendous publicity. Fu'ad Husayn, a Jordanian who has written a book on Abu Musab al-Zarqawi, notes that the U.S. response to Zarqawi's brutal violence in Iraq made him a hero. The United States made Zarqawi its public enemy number one and singled him out above other jihadists in its official statements. As a result, Husayn reports, "Every Arab and Muslim who wished to go to Iraq for jihad wanted to join al-Zarqawi" (Husayn, 2005, p. 4).

International terrorism also may lead to pressure on the local government to overreact. A government, particularly an autocratic one, may be able to ignore provocations involving attacks on its own citizens, but it cannot ignore those involving attacks on citizens of a major power.

Ironically, for purposes of creating an identity, proto-insurgents can succeed by failing to be as violent as they seek to be. Terrorist attacks or guerrilla-style strikes that fail may still draw the ire of the state, yet because little blood was spilled, the state does not achieve the same degree of legitimacy for its crackdown. Thus the proto-insurgents draw benefits from a harsh state response without alienating people by their own violence.

---

[3] See Byman (1998) for a review.

As discussed further below, violence can backfire on the proto-insurgents, despite all these potential benefits. Simply put, few people support violence. Thus proto-insurgents face the dilemma of having to decide whether and how much to target civilian populations. They seek to provoke a reaction from a government or from rival groups but not to alienate their constituencies. The Egyptian groups discussed in Appendix A, for example, alienated middle-class supporters in particular through what was perceived as senseless and indiscriminate violence. Violence is more likely to gain support if it is committed in response to a state's brutality or if it is directed at foreigners.

CHAPTER FOUR

The Proto-Insurgent's Tasks

To gain the size and capabilities of an insurgency, a terrorist group or other would-be insurgent movement must take several steps: It must create an identity, attach this identity to a cause that has widespread appeal, manage relations with rivals, find or foster a sanctuary, and address the issue of outside state support. These steps must usually be taken simultaneously and incrementally; success in one often contributes to success in another. Indicators for when a proto-insurgency is succeeding are discussed in Appendix C.

Identity Creation

First, proto-insurgents must create a politically relevant identity—a surprisingly difficult task. Individuals have multiple identities. The typical American may have a national identity (American); an ethnic one (Irish-American, say); a political one (Republican, Democrat, or myriad small third parties); a regional one (he loves NY, or he doesn't mess with Texas); and so on, all with no apparent conflict.

In time of war, it is the national identity that typically attracts the individual's loyalty—and that is what the proto-insurgent seeks to change. Rather than being loyal to Lebanon, for example, the proto-insurgent wants to foster loyalty to the Lebanese Shi'a community and the idea of an Islamic revolution and thus supports its champion Hizballah; in Turkey, one is encouraged to be a Kurd rather than being loyal to Turkey; and so on. As this suggests, identity is often created in opposition to the state. To be a Tamil is not to be Sri Lankan.

But the state is not the only rival identity. Politically, a Palestinian nationalist movement competes not only with the Israeli state (an easy match), but also with pan-Arab groups, Islamist organizations, and tribal identities, among others. For a young Fatah member today, to be a Palestinian is not only not to be Israeli, but also not to be Arab or Muslim, in a political sense.

Consider Hizballah's rise (described further in Appendix A). Before the movement emerged in the early 1980s, Lebanese Shi'a had fought on behalf of the Lebanese state, myriad Palestinian groups,[1] and various leftist movements. Over time, the dominant Shi'a group was the Amal movement, which espoused a vision for the Shi'a of communal equality with other

[1] Imad Mugniyeh, who later became Hizballah's chief terrorist operative, began his career as a Fatah operative (Jaber, 1997, p. 115).

Lebanese groups. When Hizballah began to emerge and push for an identity based on revolutionary Shi'a Islam as championed by Iran's Ayatollah Khomeini, it first had to compete with various Shi'a groups with similar identities, such as the Lebanese Da'wa and the Islamic Students Union. When these groups became incorporated into it, Hizballah took on Amal and other rivals, becoming the dominant group over time. At each stage in its rise, Hizballah had to compete with rivals on identity grounds.

Not surprisingly, common demands of proto-insurgent groups focus around identity issues, such as language, respect for a culture, or other ways of differentiating one community from another. Proto-insurgents seek to create a nation first, and then to capture a state to represent it (Byman, 1998, pp. 154–155). As Arline and William McCord argue, "For a separatist movement to emerge, people must first be convinced that they share something in common against an enemy" (McCord and McCord, 1979, p. 427).

When a proto-insurgency is developing, culture becomes intensely political. A snatch of lines from the Kurdish epic *Mem-u-Zin* or attending a play that glorifies a Muslim (vice a Palestinian or Arab) perspective on Israel both are part of this process. The IG and EIJ both targeted secular intellectuals in Egypt at the start of their campaign against the government in the 1990s, not only out of a hatred of the intellectuals' values, but also because the identity they represented—a nationalistic one, with Arabness and Egyptianness at the center—was the movement the Islamists opposed.

A rival identity cannot simply be created from scratch. Some identities lack the necessary salience, particularly if they cannot draw on language or a strong historical identity. At times, rival identities are too strong: It would be hard to create a politically dominant "Virginian" identity in the United States, even though such state-level identities spawned a civil war more than a century ago. Fearon and Laitin note that the overwhelming majority of the world's ethnic groups live side by side, if not always arm in arm, with little violence, in part because they do not see their identities as inherently mutually hostile (Fearon and Laitin, 1996).

Leaders can be tremendously important in creating an identity. Most analyses of insurgency leadership focus on tactical skill or, at times, strategic judgment. However, a leader's ability to inspire a potential group to think of itself as a people is also vital. In this sense, Yasir Arafat, who is often derided for his corrupt management style and his unwillingness to confront hardliners among the Palestinians in order to forge a peace settlement with Israel, was a remarkably successful leader. When Arafat first founded his group, Fatah, in the late 1950s, the Palestinians were divided internally along clan lines, and many felt more Arab or Muslim (or even Jordanian) than Palestinian, people of a nation without a state. By 1994, when Arafat returned as the leader of the Palestinian Authority, the Palestinian identity was one of the strongest in the Arab world.

Finding an Appealing Cause

An identity alone is not enough. The identity must be linked to a popular grievance that inspires individuals to join a revolt against the government. One can feel oneself to be a Basque or a Kurd or a Palestinian, but this does not necessarily mean that one feels the government (or, as the movement would have it, an occupying power) is illegitimate—or even if it is, that action must be taken. A state crackdown can create outrage, but this anger might only help peaceful activists or rival organizations if they are politically dominant or it might be diverted into uncoordinated and politically useless spontaneous demonstrations. Thus, insurgencies need a cause—or, ideally, several causes—that has wide appeal. With a popular cause, would-be insurgents can raise recruits, attract funding, and build their organization. Without a cause, they are simply bandits, fighting for money or to glorify violence in the name of an obscure or, at most, romantic identity.

Almost every proto-insurgent has a cause, but many do not have one with wide appeal. Proto-insurgents often begin with a cause wrapped up in their identity that appeals to only a small number of people. For example, diehard Marxist students might form a group advocating socialist revolution, but only when this idea is linked to land reform does it gain appeal outside narrow circles. Similarly, Egyptian Islamists long championed a variety of revolutionary beliefs, but these gained far more appeal when they were linked to problems such as corruption and the negative effects of Westernization.

Gaining a cause with wide appeal is harder than it sounds. A cause must resonate among a large number of people—there is no appeal in calling for the rule of the proletariat in countries with no working class. Even if the audience exists in theory (e.g., a large number of Muslims for an Islamist proto-insurgency), it may reject the cause proffered. Opposition leaders have long tried to raise the banner of Islam in the Arab world, but in modern times, it was not until the 1970s that they began to achieve success.

Fearon and Laitin find that political grievances such as a lack of civil liberties or ethnic discrimination do *not* make a country more likely to suffer from civil violence. In addition, they find that so-called plural societies, where multiple ethnic groups live and compete for power, are not more prone to conflict (Fearon and Laitin, 2003, pp. 75–78). Grievances are part of the conflict, but Fearon and Laitin argue that the conflict often produces and exacerbates them rather than the other way around, claiming that "it seems quite clear that intense grievances *are produced by* civil war—indeed, this is often a central objective of rebel strategy" (Fearon and Laitin, 2003).[2]

In this analysis, *opportunity* is the key variable determining whether civil violence will occur. Many actors engage in civil war seeking opportunities for personal gain rather than to redress grievances.[3] The resources for rebellion, such as weapons, are often cheap, and in many places the "labor market" offers access to many young men who will work for the opportunity to pillage, as they have few other opportunities that pay well. The ability to gain money from a diaspora or to extort money from exporters also appears to be important. Other opportunities include terrain that favors guerrilla war and a government that is too weak to enforce order (Collier, Hoeffler, and Sambanis, 2005, pp. 7–20).

Sequencing is a problem. Proto-insurgencies tend to attract highly committed ideologues at first. As Hoffman notes, terrorists are "violent intellectuals" (Hoffman, 2006, p. 43). Such individuals often prefer ideological purity over the sacrifices necessary to make their beliefs

---

[2] See also Collier, Hoeffler, and Sambanis (2005, p. 18).

[3] See in particular Collier (2000) and Collier and Hoeffler (2004).

more appealing. For example, some Algerian Islamists who were affiliated with the Armed Islamic Group indiscriminately slaughtered Muslim civilians during a civil war in the 1990s, arguing that because the civilians did not openly resist the government they were apostates and deserved to be executed. Not surprisingly, this killing proved highly unpopular, but the true believers refused to make the necessary compromises.[4]

Making sequencing even more difficult, the government should not be able to co-opt the issue. Galula, in his classic work *Counterinsurgency Warfare,* notes that the Philippine government seized the banner of land reform from the insurgents it faced, effectively depriving them of their most popular issue (Galula, 1964, pp. 20–22). Similarly, Egypt, Saudi Arabia, Jordan, and other states in the Arab world have tried to seize the banner of Islam from their critics by portraying their regimes as defenders of the faith.

Creating a popular cause, particularly one that cannot be co-opted, is often beyond the group's direct control. Economic disparities, resentment toward a foreign government, and other grievances are difficult to manufacture from whole cloth.

Islamists have an advantage in this regard, albeit a limited one. It is easy to accuse a regime of not being sufficiently pious: There is no accepted standard, and the demands of power and the perceived shortcomings of individuals inevitably lead regimes to stray from the mark. Yet even here, the cause cannot simply be manufactured. It was far easier to accuse the late Saudi King Fahd, who spent much of his youth carousing and whoring, of being un-Islamic than it is to level the same charges against his more pious and austere successor, King Abdullah.

Nationalism is perhaps the most potent cause to harness. At its essence, nationalism as a political movement tries to pit people of one identifiable group—defined by a shared language, race, culture, history, or other identifiable characteristic—against others to attain political control.[5] Nationalism takes various guises: Anticolonialism was one of the most common and successful variants in recent history, leading several dozen insurgencies to success against French, British, Portuguese, Dutch, and other colonial powers. Ethnic separatism is a variant of nationalism with similar dynamics, in essence, transforming a particular ethnic group into a nation with its own political identity. In recent years, Islamists have glorified the *umma,* the community of Muslims, in their attempt to create a distinct identity according to religion, in contrast to an identity based on language, defying traditional nationalism, which draws boundaries among believers based on geographic location, historical experience, and language. But such attempts frequently involve a form of traditional nationalism, pitting one group (e.g., Muslims) against outside rivals. Nevertheless, as Graham Fuller contends, Islamist struggles are far more likely to succeed when they can harness nationalism, as they have done in Chechnya, Kashmir, Palestine, and elsewhere (Fuller, 2004, p. 126).

If the government is weak, the cause need not appeal widely to produce an effective insurgency. In insurgency and counterinsurgency, strength is always relative: Hizballah needed to be formidable because it faced the Israel Defense Forces (IDF), one of the world's premier militar-

[4] Many claim that at least some of this slaughter was carried out by the government of Algeria (see Samraoui, 2003). I would like to thank Jeremy Shapiro for bringing this to my attention.

[5] The literature on nationalism is vast. For a sample, see Anderson (1983); Greenfeld (2003); Connor (1994); and Van Evera (1994).

ies. But even weak organizations can pose a threat to stability if the government they are fighting is weak. Many insurgencies consist of only several hundred fighters (Fearon and Laitin, 2003). Nevertheless, the better the cause, the easier it is for the proto-insurgency to attract recruits, money, and other forms of support and deny it to the government they oppose.

The cause should not necessarily be equated with a plan for governance. Most insurgencies that seek power have at best vague plans for what they will do if they succeed. Usually, governing plans are instrumental, part of how proto-insurgents hope to make their cause more attractive and thus gain support. The proto-insurgents, of course, are free to emphasize only the benefits of their proposals without discussing the costs or difficulties inherent in them.

The Role of Social Services

Being able to create or tap into a social services network offers tremendous advantages to proto-insurgencies. On a practical level, it allows them access to legitimate institutions from which they can divert money or find employment for followers. In addition, the social services network can assist group logistics—the network that transfers supplies for a hospital or provides documentation for workers can do the same for members of the insurgent movement.

But even more important is the challenge that a social services network poses to a government. Such a network, when cast in opposition to the state, raises the question of whether a regime is providing for the needs of its citizens. In Lebanon, Hizballah has highlighted the greater efficiency of its hospitals and social services, in contrast to those of the corrupt and inefficient Lebanese state.

Relations with Rivals

Because there is competition for both identity and the cause itself, the primary foe of the proto-insurgent in the early stages is perhaps not the government, but the welter of rival organizations vying for recruits and money. Many of these organizations seek to exploit the same cause but use a different identity or platform to do so. Still others seek to exploit the same cause but reject the need for violence. These peaceful groups often start at a relative advantage, as violence is both risky and unpopular, leaving the proto-insurgents in a difficult position. Were they unified, all these violent and peaceful groups might make a strong insurgent movement. Divided, they pose at most a limited threat.

Not surprisingly, an examination of the early days of many insurgencies suggests that their energies are consumed by fights within their community. The LTTE branded many of its rivals as traitors and spent much of its time attacking them in the movement's early stages. Palestinian groups have targeted more-moderate Palestinians for violence. Fatah spent much of its early stage trying to gain preeminence among the myriad organizations that sprang up within the Palestinian refugee community.

At times, the proto-insurgents may have a relationship with an open and legal political group. Sinn Fein was supposedly independent of the PIRA, but it was an open secret that its leadership was subordinate to the militant side of the movement. The Basque separatist Euskadi Ta Askatasuna (ETA) worked with the political group Herri Batasuna. For example, ETA would hold a demonstration calling for Basque independence that the police would break up when stone throwing began. In retaliation, ETA would then kill a policeman, sparking a cycle of reprisals. To commemorate the resulting loss of "patriots," Herri Batasuna would then hold another demonstration.

Clever proto-insurgents stir up rival communities as well, fostering communal tension to further prove the point. A brief Muslim uprising in Algeria in 1945 led to the deaths of 103 Europeans, but the European population's response led to thousands of Muslim deaths—possibly as many as 45,000 (Crenshaw, 1995, p. 479). Many Muslim leaders opposed the initial violence, but the massacres led them to change their positions. Not surprisingly, proto-insurgents at times also seek to change demographics. Population displacement serves an important service for insurgency, discrediting governments and increasing the climate of radicalization. Identity politics thus creates refugees, as well as strategic goals and the depredations that usually accompany war (Weiner, 1996; Posen, 1996).

Hiding and Sanctuary

Even as proto-insurgents strive to create a cause and wrest it away from their competition, they also need a respite from police, intelligence, and military services.

Because proto-insurgencies are small, the arrest of a few dozen activists can shatter a group. The botched attack by one member of the November 17 Organization led to the arrest of much of the group, effectively ending it as a threat. In 1996, EIJ suffered a massive blow when Egyptian security forces arrested hundreds of operatives. The result of this security failure was the collapse of the group's network within Egypt.

Many groups thus find that a sanctuary or "no go" zone is often essential for their survival. If proto-insurgents cannot hide and their cause is not popular, the government can easily convince people to denounce them. Even if the cause is popular, many citizens denounce the proto-insurgents out of fear of the much stronger regime. In a sanctuary, therefore, a group can plot, recruit, proselytize, contact supporters around the world, raise money, and—perhaps most important—enjoy a respite from the enemy regime's counterterrorism and COIN efforts that enables the members to escape from the constant stress that characterizes life underground.[6] As the Egyptian government gained the upper hand over EIJ and the IG, it steadily shut down safe areas in upper Egypt and sympathetic neighborhoods in Egyptian urban areas. Israel, in contrast, was never able to shut down Hizballah sanctuaries in the Bekaa Valley, Beirut, and, after a while, south Lebanon.

Geography plays a vital role for many groups. If insurgents can hide in mountains, jungle, or remote parts of the country where government penetration is weak, their own counterintelligence network does not need to be very robust. Rural areas can have particular advantages for defense against the police and intelligence forces of a state. In rural areas, personal net-

[6] For an argument on the importance of such stress, see Hoffman (2002a, p. 88).

works are strong, and as a result, it is far harder for denunciation to be anonymous (Fearon and Laitin, 2003). But as with so many factors related to insurgency, advantages in one area (in this case, hindering government intelligence-gathering) are offset in others. Guerrillas who are isolated in remote jungles and mountains often attract little attention and have little impact. Urban areas are riskier for insurgents if the government has strong intelligence and a large police or military presence. However, indiscriminate violence or arrests in urban areas are riskier for governments, as they can quickly lead to widespread anger at the regime as well as media attention to the government's mistakes.

The Role of Outside Support

State support offers numerous advantages to groups seeking to become insurgencies. One of the most important forms of assistance a state can offer would-be insurgents is a safe haven. During the 1980s and 1990s, Iraqi Shi'a, for example, were able to organize themselves and receive essential military training in Iran—activities that would have been impossible in Iraq under Saddam Husayn's tightly controlled regime.[7]

Refugee camps can also function as a form of safe haven—one that usually requires the support, or at least the acquiescence, of the host state. Refugee camps created by international organizations may also prove sanctuaries that are politically difficult for government forces to attack. In such cases, the camps are liable to become safe places for the combatants' dependents, bases for organizing, and sources of food and shelter for the fighters. The Taliban emerged from the squalid refugee camps in Pakistan; the Palestine Liberation Organization (PLO) drew its fighters from Palestinian refugee camps in Jordan, Lebanon, and elsewhere in the Arab world; and Hamas today draws heavily on refugee camps in the Gaza Strip for recruits.

State support usually makes a group more capable, enabling it to resist government counterterrorism and COIN campaigns and thus survive to conduct more attacks or engage in political mobilization. Most small groups of would-be insurgents are highly motivated, but they make amateurish mistakes, such as allowing too many people to know their plans or talking freely on the telephone. Iran has taught its proxies important operational security techniques that have enabled them to better hide from the governments they jointly oppose. Other groups are simply short of money and equipment. Libya's provision of arms to the PIRA, for example, gave the group more guns than gunmen and a huge supply of the plastic explosive semtex, a tactical advance for a movement that had always struggled to acquire arms and explosives.

Often the biggest impact of state support is the hindrance of a target government's campaign against a terrorist group. State aid can help a proto-insurgency endure, and endurance is key to success. As Hoffman has remarked, terrorists win by not losing (Hoffman, 2002b, p. 311). When a group enjoys a sanctuary and organizational assistance from a foreign country,

[7] At times, neighboring states provide insurgents or terrorists with a haven simply because they are incapable of ousting the rebels themselves. The Islamic Movement of Uzbekistan (IMU) has enjoyed a de facto haven in the Ferghana Valley, where the borders of Uzbekistan, Tajikistan, and Kyrgyzstan meet, because none of these regimes can police this area well.

a knockout blow is far more difficult for the government it opposes. Often the leader resides in a foreign country, so the arrest of operatives in the targeted country, by itself, will not lead to his own arrest. Turkish army and security forces regularly devastated the Kurdistan Workers' Party (PKK) cadre in Turkey and scored numerous operational successes. However, the PKK's leader, Abdullah Ocalan, lived in Syria, from which he could serve as a rallying point and direct operations despite COIN successes. Other PKK leaders also lived abroad, enabling the PKK's core to stay intact. Military action or arrests of proto-insurgent leaders risk an international clash, something most governments shy away from.

State support helps groups overcome logistical difficulties, one of the most serious challenges facing them. Many insurgencies are supported by a vast apparatus of people who procure supplies, offer military training, provide false documents, run safe houses, and take care of the families of activists who die or go to prison. Much of this can be done in the territory of a foreign sponsor, making it hard for the state to shut the activity down.

State support also hinders intelligence-gathering. When a group is small, its primary—or perhaps only—form of protection is secrecy (McCormick and Owen, 2000, p. 175). If group members' names and locations are revealed, regime security forces are often able to quickly arrest or kill them. In general, it is far easier to place spies and informants in areas where the government controls territory. As discussed further in Appendix A, after the 1967 war, Israel's domestic intelligence service established a network of literally thousands of informers throughout the West Bank and Gaza Strip that enabled Israel to disrupt many terrorist attacks and prevent anti-Israeli groups from forming in the first place. In contrast, Israel's efforts met with far less success in Lebanon when it occupied territory there after its 1982 invasion. A major Israeli problem was that Israel did not control Lebanese territory the way it controlled the West Bank and Gaza. As a result, Hizballah had far better local intelligence than the Israelis had and was effectively able to coerce or intimidate those Shi'a who might otherwise not have supported its efforts.

State support also legitimizes a proto-insurgent group and makes government delegitimization efforts almost impossible. As noted above, the PLO was widely recognized as a de facto government in the 1970s, despite the limited success of its operations against Israel. Even those governments that did not support the PLO feared that a rejection of the cause or too close an embrace would anger the organization's Arab state supporters, a cost few would pay (Rubin, 1994, p. 128). Support need not be as widespread as that enjoyed by the PLO. A group's potential followers may care more about the opinion of a few supporting states than about that of the more ineffable "world community." Lebanese Shi'a, for example, cared more about the opinion of the revolutionary Shi'a government in Iran than they did about the opinion of the world's major Western powers.

The rewards that flow from state backing often come from a poisoned chalice. Even the most supportive and ideologically sympathetic regimes have their own distinct national interests and domestic politics, making them unlikely to completely embrace the proto-insurgents' agenda. And many states see the terrorists and would-be insurgents as proxies who can be discarded according to the needs of the moment.

One of the biggest costs to a proto-insurgent group that gains state support is a decline in its freedom of action. States often act as a brake on their proxies, hoping to use them to bleed a foe or keep a conflict bubbling, but not wanting to risk greater escalation into an outright war. States rarely support proto-insurgents when they have military superiority over their foe. Moreover, states at times use terrorists as a deterrent, preferring to keep them strong but to contain their violence. Iran, for example, has cased numerous U.S. embassies and might strike in response to what it sees as a U.S. provocation. Many states (correctly) do not trust their erstwhile proxies, fearing that their alliance of convenience could turn against them.

A deeper problem involves a group's legitimacy. Many proto-insurgent groups fight in the name of liberation, either national or religious. The Mujahedin-e Khalq (MEK), a terrorist group that opposes the clerical regime in Tehran, lost any legitimacy it had when it began to conduct operations out of Iraq during the Iran-Iraq war. As Wilfried Buchta contends, "The large majority of Iranians inside and outside the country reject the MEK because of its support for Baghdad during the Iran-Iraq War and its continuing alliance with Saddam. As a result, it has only a small, dwindling power base in Iran" (Buchta, 2001, p. 116). Because nationalism is such a potent mobilizing tool, groups that allow governments to play this card have lost a tremendous advantage.

States providing support also want groups to be limited in their scope and power, as too much consolidation would threaten the sponsor's control of the overall cause. Pakistan supported the Islamic rivals of the Jammu and Kashmir Liberation Front (JKLF) even though the JKLF was initially stronger, because Islamabad did not support the JKLF's agenda. Syria backed numerous Palestinian challengers to Yasir Arafat's leadership of the PLO, preferring weak proxies under Damascus' thumb over a stronger but more independent movement.

At times, a state may even crack down on the violent group it once supported, effectively turning a haven into a prison. Until it was expelled from Libya in 1999, the Abu Nidal Organization enjoyed a haven there. Yet in the previous few years, Qaddafi had clamped down on the group, preventing it from conducting operations that might damage his attempts at a rapprochement with the United States. Such fickle support can be bloody. A PLO intelligence chief estimated that the Arab states—not Israel—inflicted three-quarters of the casualties the organization has suffered in its history (Rubin, 1994, p. 124).

State support can at times even weaken a proto-insurgency's operations and appeal. State support interrupts the evolutionary process that successful insurgents go through as they learn how to evade or weather a state's counterterrorism and COIN responses. British counterterrorism officials often referred to the "Paddy factor" when assessing PIRA operations, noting that many failed due to bungling and a lack of training. The PIRA, of course, over time became far more deadly and was eventually one of the world's most skilled terrorist groups. If it had relied heavily on state support, it might not have been forced to learn the operational skills to prosper on its own. Perhaps most important, a state patron can lead a terrorist group to lose touch with its most important constituency, the people it seeks to lead. The proto-insurgency may develop a political and military strategy that alienates the people, making them less likely to support the organization in the long term.

Nonstate actors, particularly diasporas, but also extremely broad transnational terrorist movements such as the current *salafi* jihadist movement, can also fulfill some of the functions of state sponsors. When al-Qa'ida had a haven in Afghanistan, it trained literally thousands of fighters and indoctrinated them with its broader agenda. These fighters later formed the backbone of many local groups and led several of them, such as those in Algeria and Kashmir, to embrace goals beyond their regional aspirations.

Today, the greatest contribution of these nonstate actors is usually financial. Al-Qa'ida has channeled tens of millions of dollars to various proto-insurgencies around the world. It helped seed the Islamic Movement of Uzbekistan (IMU) and played a role in the creation of the Abu Sayyaf Group, among others. In addition, it aided stronger movements and other terrorist and insurgent groups. The money it provided helped its beneficiaries gain an advantage over their rivals. Other groups, including the LTTE and the Lebanese Hizballah, have drawn heavily on support from diaspora communities. Because proto-insurgencies facing weak governments do not need to be very strong, relatively little financial support or the influx of even a few dozen seasoned fighters can transform a group and make it far more likely to become a successful insurgency.

Like state sponsors, these nonstate actors bring risks as well as benefits. They may have agendas that are disconnected from the immediate needs of the group. In particular, the transnational agenda of al-Qa'ida may at times hinder a group's local goals—al-Qa'ida may push a local group to attack the United States or otherwise embrace goals that lead to more support for the opposing regime.

CHAPTER FIVE

The Role of the State

The success or failure of a proto-insurgent movement depends only in part (at times only in small part) on its own campaign. The reaction of the state is often the most important factor in the movement's overall success or failure.

States have tremendous potential influence over would-be insurgents' ability to create an identity and organize. They can disrupt organizations through various forms of policing and repression and can co-opt potential leaders and make them allies of the state. Saudi Arabia's post-2003 campaign against would-be insurgents in the Kingdom is instructive. On the one hand, the Saudi security services began a massive effort to track down and arrest or kill terrorists there. On the other hand, the regime reached out an olive branch to an array of Islamist opposition figures, particularly religious leaders whom the jihadists admired. When these leaders issued pro-regime statements, the appeal of the jihadists to would-be recruits diminished considerably.[1]

In addition to co-opting and repressing, states can also divide the identity that proto-insurgents wish to put forward. When Israel was created in 1948, its population was between 15 and 20 percent Palestinian—a troubling fact, given the civil war between Arabs and Jews that attended Israel's birth. The Jewish state used a series of administrative steps and inducements to divide this bloc, creating incentives and means for Druze, Christian, and Bedouin Palestinians to differentiate themselves from their fellow Palestinians. Israel also tried to play up family and tribal divisions. The result was a weak and divided Palestinian community that, even after Israel took control of the West Bank and Gaza in 1967, never posed a serious threat to the state's security.

When Does Control Work?

One key question is whether a state can effectively use what political scientist Ian Lustick has dubbed "control" methods—the use of various forms of coercion and efforts to inhibit organizations—to keep the peace (Lustick, 1979, 1980). States are in a bind when it comes to force. If they crack down too hard, they risk alienating the population and creating support for organizations where none previously existed. Failure to crack down, however, can decrease

[1] An excellent review of divisions within the religious communities in Saudi Arabia can be found in International Crisis Group (2005).

confidence in the state and make it easier for proto-insurgent groups to mobilize would-be followers, since they need not fear that they will be arrested. In addition, a weak crackdown may lead rival communities to act on their own. If a group is singled out for repression because of its ethnicity, religion, or other features, the salience of that identity increases.

Similarly, state repression can politicize a community that previously was wary of politics. State measures to crush a group after an attack may involve detentions, cordon-and-search operations, increased monitoring, and other measures that can increase a population's hostility. In using such measures, the state validates the cause the proto-insurgents espouse. The Sri Lankan government's crackdown on the Tamil community in response to LTTE attacks lent credence to the LTTE's argument that the state acted on behalf of only the Sinhalese. Aggressive British government sweeps of pro-PIRA neighborhoods in Northern Ireland and tough tactics in putting down riots fostered sympathy for the PIRA (Bell, 2003).

Governments often confuse opposition political activists with those who use violence or confuse one group of extremists with another at the proto-insurgency stage. During the early stages of "the Troubles" in Northern Ireland, the militants went underground after major terrorist attacks, in anticipation of the government response. As a result, the British authorities often detained old-time activists, thereby removing relative moderates from the scene (Bell, 2003, pp. 375–381).

Governments may use attacks by extremists as an excuse to weaken moderate groups that oppose them. Egypt, for example, repeatedly arrested members of the less-violent Muslim Brotherhood in crackdowns triggered by terrorist acts of groups such as EIJ. At times these arrests are made in error, when the government does not know the community well enough to sort out the radicals from the moderates within the opposition. But often the apparent confusion masks a deliberate decision to use counterterrorism and COIN to clamp down on dissent in general.

Some regimes cannot implement control measures, while others can capitalize on circumstances that increase the effectiveness of control. The resources of a would-be insurgent group, its level of outside backing, its access to arms, its social structure, and the type of regime it opposes are among the many factors that affect the effectiveness of control. Of course, an important point not to be overlooked is the fact that governments with a high degree of popular support and skilled police and intelligence services are more likely to implement control policies more effectively.

Israeli Arabs for many years lived barely above subsistence levels. The *fellah* who might support violent resistance to Israeli rule under the military government, for example, would find his land, as well as that of his family, confiscated. Even tacit support might lead the government to cut off an individual from the land, which often meant the difference between economic prosperity and survival. As time went on and the Israeli Arab population became wealthier, however, the impact of economic punishments decreased.

The need for at least minimal wealth in order to resist authority is common for all forms of resistance. As Eric Wolf notes, "A rebellion cannot start from a situation of complete impotence; the powerless are easy victims" (Wolf, 1999, p. 290). At the same time, greater levels of wealth are a disincentive for rebellion. When individuals have too much to lose from change, they are often reluctant to pursue it.

Civil society (social clubs, religious organizations, unions, sporting groups, and other associations that are independent of the state) defines an important resource of a group, and control is more effective when civil society is weak. The communal organizations that make for a vibrant civil society also provide a ready base of organization for political action. Thus, many governments seek to shatter civil society. In most Arab countries, nearly all forms of organization, including innocuous groups such as soccer clubs, are carefully monitored.

A group's access to arms is another key resource, although acquiring arms is less of a challenge today than it was in the past. Even poor groups can acquire light machine guns, mortars, and other portable and cheap weapons that can prove exceptionally deadly. As the Carnegie Commission on Preventing Deadly Conflict notes, "The worldwide accessibility of vast numbers of lethal conventional weapons and ammunition makes it possible for quite small groups to marshal formidable fire power" (Carnegie Commission on Preventing Deadly Conflict, 1997, p. 16). Thus, controlling violence is far more difficult today than it was in the past, when governments generally outgunned potential rivals.

Control is also far harder to implement when a group has outside support. Much of the success of control depends upon expectations. Governments must convince individuals that organization for violence (or for political activity in general) is fruitless and possibly dangerous. When leaders and larger cadres have a haven outside a country's borders in which to gather, train, and organize with impunity, creating an expectation of defeat is far more difficult. Not surprisingly, groups that have repeatedly received support from neighboring governments (e.g., the Kurds) have proven far harder to control.

Governments have trouble implementing control when the terrain makes movement difficult and aids concealment. If roads and other forms of transportation are well-established, governments can move their forces quickly and can respond rapidly to flare-ups in violence. But if the transportation grid is poor, as it is in much of Africa, government forces often arrive too late to have an impact. This problem becomes particularly acute in communities located in jungles, mountains, swamps, or other inaccessible terrain, where it is easy for groups to act and organize with relative impunity. Inaccessible terrain increases the plausibility of revolt and decreases the credibility of government promises of security.

In addition to group resources in general, a group's social structure has a tremendous impact on the effectiveness of government control. Control is far more effective when there are divisions within the group that can be exploited, divisions that in turn depend in part on the strength of the group's identity. The Israeli government, for example, found it relatively easy to separate Israeli Arab Christians, Druze, and Bedouins from the mainstream Israeli Arab community, because preexisting tensions had prevented a strong, overarching identity from jelling. In general, groups that are clan-based or rely on only a few established leaders have proven easier to control because they have a readily identifiable organization structure that a government can target. When the leaders are controlled, the group as a whole generally follows. Thus, colonial governments gained tremendous leverage over various tribal groups by co-opting and controlling their leaders.

Subethnic divisions, of course, are not the only sources of division. Class, region, religion, language, and ideology often present important cleavages that enable governments to divide groups and keep them weak. Not surprisingly, groups constantly strive to unite their potential followers. Much of their violence is aimed at increasing internal support and eliminating alternative sources of political identity (Byman, 1998, pp. 157–158).

The ideal allies for a government implementing control are, in fact, nonviolent members of the community the would-be insurgents seek to mobilize. Strong moderate forces can be interlocutors to the community in general and an alternative for political action that does not involve violence. If moderates side with the government, they can provide superb intelligence on radical activities. Governments in India and Spain have worked with moderates against radicals in their own groups.[2]

In essence, governments must break the link between mainstream groups and those that would use violence. Political movements often serve as fronts for terrorist and guerrilla groups or work with them side by side. They provide social services, run businesses, and provide a legitimate face for fundraising and political activities. Moreover, they offer a means of recruiting new members. If regimes can infiltrate—or, better yet, cooperate with—mainstream groups, they are often able to gain information on radical activities and turn potential militants away from violence.[3]

Good intelligence is essential when implementing control. If the government uses force against otherwise passive individuals, it risks turning them into rebels. The military government in Israel encouraged positive leaders while it was ruthlessly suppressing those considering the use of violence, in part by gathering extensive intelligence on the Israeli Arab community. The Iraqi Baath, on the other hand, used violence indiscriminately, thus encouraging passive individuals to rise up. Security measures should be applied consistently and, when possible, with restraint, in order to prevent a backlash that will cause governments to lose the support of the broader, apolitical population (Byman, 1998, p. 162).

Control is easier to implement if the country's security forces are drawn from all of its communities. The more the state is an agent of change or ethnic hegemony, the more control is required to keep the peace—and the more it will be resented. In Israel, the state was the agent of the dominant Jewish community and was resented as a result. Iraq's experience is the most telling: The state was both an agent of assimilation and controlled by one communal group. The failure of the Baath to keep the peace suggests the limits of control under such trying conditions.

Regimes that are respectful of civil liberties find control measures harder to implement. Democracies around the world wrestle with the question of how much surveillance is acceptable. Even countries where respect for civil liberties is entrenched, such as the United States, allow the investigation of groups that might be using or preparing to use violence (Shulsky and Schmitt, 2002, pp. 148–151). Yet, as Shulsky notes, this creates a Catch-22 regarding democratic restrictions and surveillance: "One cannot know about the support group's additional activities because one may not look, and one may not look as long as one does not know" (Shulsky and Schmitt, 2002, p. 155).

---

[2] Working with moderates encourages in-group policing, as described in Fearon and Laitin (1996).

[3] For a non-ethnic example of this, see Della Porta (1995, pp. 126–127).

CHAPTER SIX

Defeating Proto-Insurgencies

One of the most important factors in combating proto-insurgencies is understanding when they are growing into more formidable foes.

Appendix C provides indicators for analysts monitoring potential insurgencies. It examines different measures of a group's size and composition, its success in creating an identity, its use of violence, ties to other groups, links to state sponsors, creation of a sanctuary, and other vital parts of insurgent success. As the discussion makes clear, it is particularly important to recognize the many dimensions of what constitutes an insurgency—simply monitoring attacks or size is not enough. It is essential for many indicators to be monitored (and refined, with new ones added as they emerge).

Governments opposing proto-insurgencies must recognize the proto-insurgents' many weaknesses. Frequently, as noted earlier, both the identity proffered and the cause championed have little support among the public, especially in the initial stages of a proto-insurgency. For example, the *salafi* jihadist credo that many would-be insurgents advocate is followed only by a small fraction of the overall movement. Indeed, not only do most Muslims reject this credo, most self-declared Islamists, and even most *salafists*, reject it as well.[1]

The violence proto-insurgents use to attract attention to their cause also alienates many would-be supporters. Gilles Kepel contends that the "savage violence" of terrorist organizations such as al-Qa'ida has worked against them. Rather than inspiring other Muslims to take up arms against the West or apostate regimes, they have disgusted their co-religionists, leading them to reject extremism. In Egypt and Algeria in the 1990s, the brutal tactics of terrorist and insurgent groups alienated potential recruits and funders, eventually leaving the groups isolated (Kepel, 2002, p. 320).[2] (This should be contrasted with the success of other violent groups, such as Hizballah and al-Qa'ida.)

Proto-insurgents also face severe organizational limits. They are rarely strong enough to openly recruit or proselytize without fear of government arrest. Communication is exceptionally difficult, but it is necessary to increase the size of the group. Increased communication in

[1] The International Crisis Group points out that *salafis* in general are highly skeptical of the legitimacy of rebellion against a Muslim government, even if the government does not follow *salafi* teachings: "Most salafists, if forced to choose between the Saudi government and Osama bin Laden, would choose the former" (International Crisis Group, 2004b, p. 4).

[2] Also, their violence often reflects a lack of grassroots support or organization.

turn makes the group more vulnerable to government disruption (McCormick and Owen, 2000, p. 176).

Globalization has lessened this weakness, although it has not eliminated it. Through the Internet, satellite television, and other modern forms of communication, leaders in exile are able to directly communicate with their followers; this is less desirable than being near the followers and able to motivate them directly, but it is a way for leaders to stay in touch. In addition, technology enables them to proselytize and raise money far more effectively, since they are able to create their own media and can directly tailor their message to would-be constituents.[3]

Terrorist and guerrilla groups also face difficulties in recruitment. Many recruit from preexisting networks: family, tribe, village, and friends (McCormick, 2003, p. 490). But groups must have size to become insurgencies, and as a result, they often have to leave relatively safe recruitment zones and go to areas where the new recruits are far more vulnerable to government penetration efforts.

Organizational competition is another weakness. Many groups try to capitalize on "the cause," resulting in a keen and sometimes bitter competition among like-minded groups, which are often one another's worst enemies rather than allies. Because many radical communities are intensely close-knit, it is often easy for government officials to gain information on a group simply by asking its rivals.

General Recommendations for Defeating Proto-Insurgencies

Perhaps the best and most efficient way to prevent proto-insurgents from gaining ground is through in-group policing.[4] Communities know their own members, particularly in tight-knit societies in the developing world. These communities thus do not present an intelligence challenge, and they enable the use of arrests or other forms of pressure with far greater discrimination.

The experience of Jewish groups under the British Mandate suggests the power of this approach. In 1944 and 1945, the Haganah, the largest moderate Jewish organization, cracked down on the more radical Irgun, believing that the Irgun's terrorist attacks on the British were hindering the political settlement that the Haganah leadership believed London was supporting—a settlement that the Haganah believed was in the interests of the Jews. The Haganah handed over perhaps 1,000 Irgun members to the British, effectively destroying the Irgun's operational capabilities in the short term.

In-group policing, of course, requires a government to work with more-moderate members of the community and often to make political concessions to them. For example, to gain the support of Irish nationalists in Northern Ireland, the British authorities had to offer them considerable power and perks. Such measures are costly for many governments, and they can be undesirable for the United States if they involve distancing government policies from those preferred by Washington. Egypt has co-opted moderate Islamists, in part by giving them

---

[3] For more on this change, see Hoffman (2006, pp. 197–228).

[4] This recommendation derives from Fearon and Laitin (1996).

control over the state's religious and judicial infrastructure and by allowing them to blast the United States in their media outlets—a "success" when looked at narrowly with regard to Cairo's efforts against the IG and EIJ, but a quite troubling one from the perspective of other concerns about the decline in the secular nature of the Egyptian state.

Working with moderates provides more than benefits for police and intelligence services. It also offers a rival path to power and influence by discrediting the standard defense of violence—i.e., that power respects only force.

States can also promote rival identities. Ideally, they can build up the national identity to increase bonds among citizens, but they can also use powerful tools such as education, control of language, and influence over the media to build a rival identity to the one the proto-insurgents support.

Another alternative is to divide the identity of the proto-insurgent movement. The Palestinian group Hamas, for example, has numerous "identity" tensions, despite in theory championing a blend of Palestinian nationalism and Islam. Hamas is divided by region (members on the West Bank versus those on the Gaza Strip) and by how much to emphasize the Islamic agenda versus purely nationalistic demands. Tribe and family also are important within Palestinian society (International Crisis Group, 2006a).

All these actions are harder for the state if the group has a haven from which to operate. In-group policing becomes almost impossible with regard to the group's leadership, although local recruits can still be targeted.

On the other hand, if the proto-insurgents accept foreign backing, it is far easier for the government to turn nationalism against them. Charges of treason are far easier to back up if a group is openly taking support from another government. However, these charges do not always stick. The Lebanese Hizballah is openly close to Iran and, to a lesser degree, Syria, but repeated efforts by its rivals to criticize these relationships have not significantly decreased the esteem in which many Lebanese, particularly Lebanese Shi'a, hold the group.

Leadership analyses must also recognize the cultural and identity sides of the equation, as well as the military side. Leaders who are vital for transforming terrorist or small guerrilla groups into insurgencies are often cultural figures or others not directly involved in violence. These individuals can play a key role in creating unity and identity where none existed before.

The clandestine nature of proto-insurgent groups and the negative effects of violence can also facilitate undermining them. Governments can at times clandestinely commit brutal attacks in a group's name or can simply allow the group to commit them unmolested to undermine the group's overall credibility. This is a brutal approach, and it involves the deliberate deaths of innocents, but it can work: In Algeria, the government infiltrated parts of the jihadist movement and encouraged it to conduct attacks on noncombatants; it then used these attacks to "prove" to audiences at home and abroad that the jihadists deserved no quarter. Because the jihadist movement had many leaders and factions, it was not able to credibly deny the attacks committed in its name.

Recommendations for the United States

The most obvious action for the United States in its COIN campaign is to anticipate the possibility of an insurgency developing before it is manifest. Many of the recommended steps are relatively low cost and easy to implement, especially when compared with fighting a full-blown insurgency.

The United States must often stay in the background when dealing with potential insurgencies. Since the best cause for insurgents to harness is nationalism, direct and open U.S. support can undercut the legitimacy of a government. The United States can, however, provide behind-the-scenes training and advisory programs, particularly if the programs are conducted outside the country. These programs should focus on improving indigenous capabilities rather than on substituting U.S. roles for them.

Building strong police forces is also important—usually much more important than aiding the military. Police typically are far better suited to defeating small groups, because they know the communities well and are trained to use force discriminately. Galula contends that the police are "the first counterinsurgent organization that has to be infiltrated and neutralized" (Galula, 1964, p. 31). Not only must the police be strong and numerous, the laws they enforce must be suited for counterinsurgency. Thus, legal reform is a vital early step in counterinsurgency (Galula, 1964, p. 31).

Historically, helping other governments improve their policing and intelligence capabilities under foreign internal defense has been bureaucratically weak in the United States. Unlike Italy, with its *Carabinieri*, or Spain, which has the *Guardia Civil*, the United States does not have a national police with a paramilitary component, making it difficult to identify an obvious bureaucratic candidate for such an important training mission. The State Department is too small to provide a massive training program, so the foreign internal defense mission would fall upon the Department of Defense, which historically has resisted it (Rosenau, 2003).

A particularly difficult problem for the United States is that while it has many programs to train allied militaries as well as a few (much smaller) ones to work with local police forces, it lacks a significant training program focused at paramilitaries beyond limited intelligence liaison efforts.[5] Creating such a program—ideally, one that combines the talents of military officers, police officials, and intelligence professionals—would be an important step forward in fighting proto-insurgencies.

Intelligence should also be redirected to focus on the conditions that can foster an insurgency as well as on ongoing insurgencies. Once an insurgency is manifest, it is far harder to defeat. Identifying would-be insurgents and assessing their chances of success would thus be a highly effective means of helping to direct resources.

---

[5] The United States has trained the constabulary in the Philippines and various forces in Vietnam (with mixed success). In general, however, these programs have not been robust or sustained.

The Risks of Success

Seemingly paradoxically, failure to foster an insurgency may make a group's overall agenda more ambitious and may lead it to engage in global terrorism. This has particularly important policy implications for the United States: Stopping a proto-insurgency in one country can increase the global terrorism threat to the United States.

For a group confronting failure, a broader agenda provides more (and often less-defended) targets to attack and a different set of grievances to exploit. These benefits can now keep a group going for years, in contrast to an era when members died or were arrested with no appreciable gain in recruits or damage to the enemy.

For example, EIJ leader Ayman al-Zawahiri had once endorsed a local agenda to the point that he declared that even the defeat of Zionism must wait until Islam triumphed in Egypt. After 1997, however, he accepted bin Laden's broader view that the root problem was America, not local regimes. Zawahiri joined with bin Laden in part because EIJ was flailing in its effort to topple the Mubarak government after what had appeared to be a promising start earlier in the 1990s. Zawahiri and his followers were chronically short of funds and were conducting few successful attacks in Egypt. Moreover, they were being hounded from place to place, particularly after Sudan expelled Islamic militants in 1996. Zawahiri's effort to set up a sanctuary in Chechnya failed, and the United States helped disrupt the EIJ base in Albania through a rendition campaign that led to the unraveling of much of its remaining network in Egypt. As one former EIJ activist reported, "Zawahiri was cornered. He had nowhere to go. He joined with bin Laden because he needed protection."[6]

Failure produced an escalation in Zawahiri's case, in part because he had few options at a local level yet had an opportunity to go global—something many groups lack. A group with a strict ethno-nationalist agenda would find it far harder to gain recruits by striking the United States or other targets not involved in the immediate struggle. Similarly, going global was not an option for militant Islamists who lost battles with the state in the 1980s, such as Syria's Muslim Brotherhood.

The failure of insurgent methods such as guerrilla war is a particularly common reason for a group to embrace terrorism or to use it as the primary form of resistance. Terrorism is often a tactic of last resort; many groups prefer to use a strategy of guerrilla warfare, believing both that it is more honorable and (more important to most of them) that it has a greater chance of succeeding over time. Some groups, such as would-be insurgents in Brazil, Uruguay, and Argentina, turned to terrorism after they failed in a guerrilla war strategy.[7] Fatah and other Palestinian groups initially sought to use guerrilla warfare but embraced terrorism when Israel destroyed their nascent organizations in the West Bank and Gaza after the 1967 war. Hizballah, in contrast, greatly reduced its own direct use of terrorism against Israel after the early 1990s as its guerrilla operations became more proficient. In Chechnya today, the Russian government has largely defeated the indigenous Chechen guerrilla movement, but the foreign Mujahedin, who primarily use terrorism, are still going strong.

---

[6] As quoted in Higgins and Cullison (2002).

[7] I would like to thank Brian Jenkins for pointing this out to me.

Final Thoughts

This paper has examined the process of insurgency development, focusing on one key stage: the state at which a terrorist group or small insurrection is able to make the transition to a full-blown insurgency. As the analysis makes clear, the challenges of dealing with proto-insurgents are related to but nevertheless distinct from those of fighting a full-blown insurgency.

It is particularly important to recognize the many dimensions of a successful insurgency. Most analyses focus on size, resources, and outside support, and some also address the issue of popularity of the insurgents' cause. Equally important, and at times more so, are issues regarding identity and group competition: The political identity that is most salient or the group among the myriad challengers that will prevail is not preordained. None of this can be separated from a government response. For almost all of these factors, the proto-insurgency's government opponent is the decisive factor in success or failure.

Finally, a word of caution is also in order. Most of the factors presented here that explain why terrorists or small bands of rebels become full-blown insurgencies can cut both ways. State support can both strengthen and weaken a group. Government COIN campaigns can crush a group, but at times they can backfire and strengthen one. Terrorist violence can attract attention and isolate moderates, but it can also alienate public support. There is no formula for success for either the proto-insurgents or the governments they oppose.

APPENDIX A

Three Cases of Proto-Insurgent Success and Failure

This appendix describes the evolution of three different groups to draw lessons on how some groups evolve from terrorist movements to successful insurgencies, while others fall short. The Lebanese Hizballah, once a small band of terrorists, grew to become one of today's most formidable insurgencies, eventually driving the IDF from Lebanon in 2000. In contrast, EIJ and the IG waged terrorist campaigns and low-level guerrilla attacks in Egypt in the 1990s, but the campaigns petered out as the decade wore on. The Palestinian group Fatah wavered between success and failure: It was unable to form an insurgency against Israel in the Palestinian territories, but it did wage guerrilla war from havens in Jordan and Lebanon for many years.

The Rise of Hizballah

Hizballah today is far different from the ragtag collection of Shi'ite fighters that emerged to shock the world in the early 1980s. Over the years, the movement has evolved militarily, becoming one of the world's most lethal guerrilla groups. It has also developed politically, becoming lionized in the Middle East and expanding the scope of its activities in Lebanon. This section focuses on the development of Hizballah from proto-insurgency to insurgency; it does not look in depth at how the movement has fared as it achieved its present status.

For most of Lebanon's modern history, the Shi'ite Muslim community from which Hizballah emerged was ignored and powerless, even though it eventually became the largest communal group in the country. Under the French colonial system, Lebanon's Maronite Christian and Sunni Muslim communities dominated the country, quarreling over who would lead it but not over whether the Shi'ites deserved a greater say. Then, beginning in the 1960s, the charismatic Shi'a cleric Musa al-Sadr challenged the traditional, politically quiescent leaders of Lebanon's Shi'ites. Sadr organized the community, demanded a greater voice in Lebanon's affairs, and, in the process, formed the Shi'ite militia Amal (the Afwaj al-Muqwama al-Lubnaniya, or Lebanese Resistance Detachments, the acronym of which means "hope" in Arabic). In 1975, in the midst of this politicization, Lebanon fell into civil war, a conflict that embroiled many of the country's religious and ethnic communities, the large Palestinian refugee presence, and neighbors such as Iraq, Israel, and especially Syria. The creation and building of militias spread quickly as communities organized for self-defense.[1]

Israel invaded Lebanon in 1982, trying to destroy the Palestinian guerrillas' haven there. The Lebanese Shi'a who lived in southern Lebanon where most of the Israeli forces were located initially welcomed the Israeli invaders because they drove out Palestinian militias (Ajami, 1986, p. 200; Jaber, 1997, p. 14). The welcome proved short-lived, however. Israel showed few signs of leaving, making the liberators appear to be future oppressors. At the same time, it tried to establish an allied (many would say puppet) regime in Beirut dominated by Maronite Christian Phalangists—a move that alienated other communities and threatened Syria and Iran. As the situation in Lebanon unraveled, the United States, France, and Italy deployed peacekeeping forces to maintain order and help ensure the smooth demobilization and departure of Palestinian fighters in Lebanon.

Nabih Berri, Sadr's rather lackluster successor (Imam Sadr himself disappeared mysteriously in 1978 on a trip to Libya), cooperated with the Lebanese government, which was now working closely with the Israelis. Several local Shi'ite factions, inspired in part by Khomeini's revolution in Iran, disenchanted with Berri's leadership, and already incensed over the Israeli presence in their country, rejected this cooperation, arguing that the regime was nothing more than an Israeli puppet. Iran and Syria encouraged this perception. Both countries sought to use the Shi'ites against Israel's interests in Lebanon, and Iran also hoped to export its revolution to Lebanon. With Syria's encouragement, Iran helped organize, arm, train, inspire, and—most important—unite the myriad Shi'ite groups, eventually leading to the formation of Hizballah.

Hizballah emerged from the hothouse of Lebanese Shi'a politics as the dominant group. It distanced itself from the Amal movement by decrying the "betrayal" of Berri for his cooperation with the Israeli-supported Maronite government. Iran and Syria, both highly hostile to that government, thus supported Hizballah over Amal.

Organizationally, Hizballah was able to co-opt many small existing groups. Prodded (and funded) by Iran, these disparate movements coalesced around Hizballah. Initially, they included the Islamic Amal movement (a splinter of the overall Amal organization), the Association of Muslim Ulema in Lebanon, the Lebanese Da'wa, and the Association of Muslim Students. Over time, the movement spread to Beirut, where it incorporated the many followers of Shaykh Fadlallah, a leading Lebanese religious scholar who at the time endorsed many of the ideas of the Iranian revolution. From there, the movement spread to the Amal stronghold of southern Lebanon, where it incorporated many local fighters who were battling the Israelis largely on their own (Ranstorp, 1997; Shapira, 1988, p. 124; Wege, 1994, p. 154; Hajjar, 2002, pp. 6–9).

A Violent Success

Hizballah quickly became the tip of the spear in the effort to expel the Americans and the Israelis. It literally exploded into America's consciousness with devastating suicide attacks (new

[1] For valuable accounts of the collapse of Lebanon into civil war, see Hiro (1992) and Hudson (1985). The best account of the role of the Palestinians is found in Brynen (1990). The steady politicization of the Shi'a is described in Norton (1987).

at the time) on the U.S. embassy in Beirut in April 1983, where 63 people died, including 17 Americans, and on the U.S. Marine Barracks in October 1983, where 241 U.S. Marines died. These attacks, along with the sense that the peacekeepers had little peace to keep, led to a rapid U.S. departure from Lebanon in February 1984. Hizballah conducted similar attacks on French and Israeli sites in the country; it also took numerous hostages, killed dissident Iranians in Europe, and bombed Jewish and Israeli targets in Argentina in 1992 and 1994.

Less noticed, but over time more important, Hizballah also began a long, bitter guerrilla war against Israel. It was initially carried out by local, relatively autonomous fighters in the south, who became more and more effective over time. Many of the tactics Hizballah initially used, such as driving truck bombs into Israeli convoys and facilities, were a mixture of terrorism and guerrilla tactics. Faced with the ferocious Hizballah attacks, Israel withdrew in June 1985 to a "security zone" in southern Lebanon—a buffer manned in part by Israel's allies, the South Lebanese Army (Jaber, 1997, pp. 16–27).[2] Hizballah kept up the pressure and eventually, 15 years later, drove Israel from Lebanese soil altogether.

Although it garnered relatively little attention in the West, perhaps Hizballah's biggest military struggle was within the Shi'a community, when it and Amal attacked each other near the end of the "War of the Camps" in 1986 and 1987, fighting in part for control of the Shi'a community.

Hizballah's size fluctuated as its struggle against Israel wore on. In the 1980s, it had perhaps 5,000 fighters under arms, several hundred of whom belonged to various front organizations, including the Revolutionary Justice Organization, the Oppressed of the Earth Organization, and Islamic Jihad. Hizballah used these fighters for its terrorist operations.[3] This cadre shrank in the 1990s but grew far more skilled and professional. By the early 1990s, Hizballah had become an elite organization, with several thousand fighters under arms, most of whom were highly trained. In its 2006 clash with Israel, Hizballah fighters proved skilled and effective, able to inflict significant casualties on the IDF.

Explaining Hizballah's Success

Hizballah's violence and superior organization enabled it to outstrip Amal and its other rivals. The suicide attacks on Western targets in particular demonstrated the movement's prowess. The spectacular nature of the attacks and the attention they received served as an excellent recruiting device. Hizballah may have avoided the backlash faced by other groups that used violence, because the targets were foreign and primarily official (military or diplomatic) in nature, in contrast to the attacks by Egyptian groups on tourists.

In developing their guerrilla organization, Hizballah's leaders realized that many of their recruits were unskilled and that large numbers of them made Israeli penetration of the group's ranks far easier. Moreover, repeated Israeli assassinations and kidnappings had demonstrated

---

[2] Many of the early attacks were carried out by south Lebanese affiliates of Amal, but over time these affiliates either joined or were overshadowed by Hizballah (Black and Morris, 1991, p. 451).

[3] Hizballah has admitted that these organizations are not separate entities (Ranstorp, 1997, p. 53) (see also Hamzeh, 1997). Other experts report that Hizballah had 5,000 fighters and 5,000 more reservists by the end of the 1980s (Wege, 1994, p. 155).

Israel's ability to gain real-time, actionable intelligence on Hizballah. Decreasing its size enabled Hizballah to increase its overall level of training and improve security. By the time of the Israeli withdrawal from Lebanon in May 2000, Hizballah had approximately 500 full-time fighters and another 1,000 part-time fighters (Norton, 2000).[4]

Hizballah's smaller size did not diminish its reach or its stature. Because Hizballah fighters were lionized by many Shi'a, it was relatively easy to gain new recruits. In addition, Hizballah worked closely with Lebanon's Shi'a religious network, giving it a built-in leadership base and ties to the community that went beyond the organization's narrow activities. Many of its recruits were bonded through kinship and regional ties, as well as through a shared ideology. Hizballah's terrorist wing in particular was able to choose among the more skilled of the organization's members, rejecting those who might in any way hinder operations.

The fighters also increasingly specialized and improved their security and logistics capabilities. The movement learned how to better use Lebanon's broken terrain, how to plan sophisticated roadside explosions, and how to coordinate small units against Israeli forces, increasing the number of casualties it inflicted. Hizballah also attacked Israeli positions with heavier and more sophisticated weapons (Jaber, 1997, pp. 37–42; Eshel, 1997). As a result of these changes, Hizballah became a formidable guerrilla force that increasingly began to impose a cost on the Israelis.

As grim as this track record is, it is important to note that the nature of Hizballah's involvement in terrorism has changed. In the 1980s, Hizballah was perhaps the world's most active terrorist organization, assassinating anti-Iranian figures, bombing a range of targets around the world as well as in Lebanon, holding hostages, and otherwise targeting noncombatants. In the 1990s, however, the movement decreased its direct involvement in terrorism, focusing more on its guerrilla struggle against Israel. In short, it became an insurgency.

Hizballah today is a social and political organization as well as a terrorist and guerrilla movement. Popular among Lebanon's Shi'ite plurality and respected by many non-Shi'ite Lebanese, Hizballah holds 23 seats in Lebanon's parliament of 128 and holds two government ministries following the 2005 elections. Hizballah runs schools and hospitals and offers relatively efficient public services, in sharp contrast to the inefficient government of Lebanon.

Hizballah's social services bolster its military and terrorist activities. The social services provide a network of sympathizers who may be called on to assist operations. The services also build the prestige of the guerrilla movement, demonstrating its ability to help the people beyond resistance. Equally important, they have helped the group discredit the government to its own advantage, as Hizballah's services involve less corruption and are more efficient than those of the state.

Over time, Hizballah developed an extremely effective electoral machine. Hizballah is respected for its efficiency and, by Lebanese standards, its limited corruption. Its electoral posters called for votes on behalf of the movement of martyrs. Hizballah also very effectively exploits its social service network (Hamzeh, 1993). As one Christian who voted for Hizballah explained to a would-be rival, "Where were you when we needed emergency snow removal and

---

[4] Other sources put the number of full-time fighters even lower, at around 300 (see Blanford, 1999).

fuel? In this village, everyone is going to vote for Hizballah."[5] Not surprisingly, Hizballah has done well in Lebanon's elections.

The Role of Iran and Syria

In different ways, Syria and Iran have played vital roles in sustaining Hizballah. For both countries, using Hizballah as a proxy allows some degree of deniability, enabling them to strike at Israel or other targets without risking a confrontation that direct military action would entail. Beyond this shared objective, however, the two states have very different approaches to Hizballah.

Syria provides Hizballah with weapons and logistical support. Most important, for many years Damascus cracked down on Hizballah's rivals while allowing Hizballah a haven in Lebanon, and today it supports the group's presence there. Syria's relationship with the group is intensely practical. Damascus has avoided direct involvement in international terrorism since 1986, but it still uses radical groups to put pressure on Israel and other countries. In essence, Syria's ties to Hizballah remind Israel that it cannot end terrorism or stabilize its border without accommodating Damascus. Syrian President Bashar al-Assad confessed this open secret, noting that Hizballah was a necessary "buffer" for Syria against Israel.

While Syria was occupying Lebanon, Damascus appeared to exercise a veto power, or at least considerable influence, over Hizballah's military operations there. As Human Rights Watch notes, "By controlling Hizballah's prime access to arms, Syria appears to hold considerable influence over Hizballah's ability to remain an active military force in the south" (Human Rights Watch, 1996, p. 22).

The recent Syrian withdrawal from Lebanon weakens Damascus' political position considerably. Even though its intelligence remains superb, its on-the-ground presence is diminished, and it now relies more on Hizballah to represent its interests than it did in the past, increasing the movement's leverage.

Iran has helped build Hizballah from the ground up and plays a major role in sustaining it on a daily basis. After Israel invaded Lebanon in 1982, Tehran seized the opportunity and deployed 1,000 Islamic Revolutionary Guard Corps (IRGC) personnel—the revolutionary vanguard of Iran's military—to Lebanon's Bekaa Valley. The number of troops later leveled out at between 300 and 500 (Norton, 2000; Shapira, 1988, p. 123). The IRGC worked with Iranian intelligence and Iranian diplomats as well as Syrian officials to create Hizballah from a motley assortment of small Shi'ite organizations. Iran helped the fledgling movement train and indoctrinate new members in the Bekaa Valley and developed a social services and fundraising network there. Iran currently provides the movement with relatively advanced weaponry.

These ties remain strong, and Iranian sponsorship of Hizballah is the primary reason for Iran consistently heading the U.S. list of state sponsors of terrorism. Hizballah still proclaims its adherence to Iran's ideology of the *velayet-e faqih* (guardianship of the jurisconsult). Tehran provides approximately $100 million per year to Hizballah, and Iranian forces train the movement and provide it with intelligence. Moreover, Hizballah operatives enjoy close ties to Iranian intelligence and the IRGC, which is connected directly with Iranian Supreme Leader Ali

[5] As quoted in Harik (1996, p. 51).

Khamenei. Hizballah's senior terrorist, Imad Mugniyeh, reportedly enjoys Iranian citizenship and regularly travels there. Hizballah proclaims its loyalty to Khameini, and he reportedly is an arbiter for group decisions. Iran is particularly influential in Hizballah's activities overseas. For example, Hizballah stopped its attacks in Europe as part of a broader Iranian decision to halt attacks there.

In exchange for the aid it provides, Iran gains a weapon against Israel and influence far beyond its borders. Because of Hizballah, Iran has defied geography and is a player in the Middle East peace process. Iran has also used Hizballah operatives to kill Iranian dissidents and to attack U.S. forces in Saudi Arabia and Germany. It uses terrorism as a form of deterrence, "casing" U.S. embassies and other facilities to give itself a response should the United States step up pressure (Pillar, 2001, p. 159). Finally, Tehran has an ideological bond with Hizballah, formed by a similar view of the role of Islam in government and historically close ties between Lebanon's and Iran's clerical establishments.

Hizballah's foreign backers are both a source of the movement's strength and a brake on its activities. Iran and Syria both use Hizballah operations to further their own foreign-policy objectives and, in the process, make the movement far more dangerous. Yet their close ties to Hizballah make them vulnerable to retaliation for the movement's trespasses, leading them to rein in the organization when they feel threatened.

Israel's Response to Hizballah's Rise

Hizballah had a particular advantage over many terrorist groups in that it did not face a functioning government on its own territory. The Lebanese government had begun to collapse by 1970, and this accelerated after the civil war broke out in 1975. By the time Hizballah emerged in the mid-1980s, Lebanon effectively had no government. Thus Hizballah had to compete with rival militias, but it did not face a state that was able to effectively divide it or offer a compelling rival identity. Equally important, the government of Lebanon was not able to coerce the Shi'a or Hizballah.

Israel fought Hizballah aggressively, but as an outside power, it lacked the ability to offset the identity politics of the group or lessen the salience of the cause it championed. Israel has used several measures to fight Hizballah, including conventional military operations against guerrilla targets; collective punishment of Lebanese civilians to undercut Hizballah's popular and government support; assassinating and kidnapping Hizballah leaders; psychological warfare and improved intelligence-gathering; pressure on Syria and Iran; using a security zone to create a buffer against attacks; and employing other groups to act as proxies in guerrilla combat.

In Hizballah's early years, all these efforts suffered because Israel lacked the necessary intelligence to shut the movement down—a problem exacerbated by the lack of Israeli control over much of the territory in Lebanon. Shabak officers in Lebanon worked with the IDF, trying to recreate the set of informers, safe houses, and other intelligence infrastructure that existed in the Palestinian territories. Shabak tried to recruit informers among the Shi'a and used harsh interrogations to dissuade potential supporters from backing Hizballah and affiliated groups (Black and Morris, 1991, pp. 395–396). Israel's efforts appear to have met with little success, although data on the specific effects of various projects are scarce. In my judgment, Israel's efforts failed in part because the negative effects of day-to-day COIN operations such as arrests, detentions, collective punishment, and harassment far outweighed the positive aspects of particular programs.

Hizballah had much better local intelligence than the Israelis had, both because of the familiarity of its operatives and because it had a vast network of sympathizers. In addition, Hizballah was effectively able to coerce or intimidate those Shi'a who might otherwise not support its efforts, while Israel's threats were less credible because the Israeli presence on the ground was not permanent. Over time, Hizballah also developed its own counterintelligence capabilities, enabling it to weed out informers and plant its own operatives in communities that cooperated with the Israelis. Moreover, in some ways Israel's attacks strengthened the movement by allowing it to demonstrate its prowess and determination. Hizballah was able to play on a sense of pseudo-nationalism, gaining status by fighting "invaders" while other Lebanese groups quarreled among themselves.

Israel's problems with Hizballah became manifest in the summer of 2006, when the group's kidnapping of two Israeli soldiers (and the killing of several others in the process) sparked a war between the group and Israel. Israeli air strikes destroyed parts of Lebanon's infrastructure, and Israel's air and ground forces clashed with Hizballah fighters in much of southern Lebanon. Although Israel inflicted considerable losses on Hizballah, it was not able to defeat the movement. Hizballah emerged bloody but proud, gaining admiration from much of the Arab and Muslim world and with its position in Lebanon possibly even strengthened.

Fatah's Mixed Record, 1959–1993

Like Hizballah, the Palestinian group Fatah, which was led by Yasir Arafat until his death in 2004, tried to evolve from a terrorist group to an insurgency. When Fatah was founded by Palestinian diaspora members in 1958 or 1959, it saw the Algerian anti-colonial struggle as its model: Fatah would use violence to launch a broader resistance movement that would expel the Israeli occupier (Rubin, 1994, p. 10). The group succeeded only in conducting guerrilla war and engaging in extensive political mobilization from exile, however—and that proved debilitating in its overall struggle.

Identity Successes

One of Arafat's greatest successes was fostering a Palestinian identity, and from there the Palestinian cause. At the time of Fatah's founding, many intellectuals and ideologues who had lived under the British Mandate in Palestine were propounding emphasized Arab nationalism and, to a lesser degree, social revolution. They found inspiration in Egyptian President Nasser, who promoted the idea of Arab unity and, with it, social revolution against conservative Arab regimes. Politically, these ideologues emphasized support for Arab and revolutionary causes over the narrower Palestinian projects that Arafat pushed. Further complicating Arafat's efforts, family and clan were often dominant within the Palestinian territories.

In practice, this identity competition posed a particular problem as Arab states sought to control the Palestinian cause. In 1964, Arab states even founded the PLO, an umbrella group meant to unite the many Palestinian factions, largely to co-opt and control the Palestinian cause. The PLO's first leader, Ahmed Shuqayri, dutifully parroted Nasser's view that Arab states would destroy Israel and restore Palestine, and that the Palestinians should thus serve the Arab cause rather than their own narrower agenda. Swept up in the fervor Nasser's charisma generated, many Palestinians initially endorsed this view, including such later leaders as George Habash (Rubin 1994, p. 9; Kimmerling and Migdal, 1994, p. 215).

Arafat and Fatah, however, were able to reverse this political identity. They believed that Palestinians must act on their own for liberation and that the Arab states should, in fact, serve the Palestinian national movement, not the other way around (Kimmerling and Migdal, 1994, p. 213).

Ironically, Israel proved to be the key to Arafat's identity success. Fatah and Arafat were able to take over the PLO after Israel's stunning victory in 1967 humiliated the Arab states and discredited Nasser's proposed solution to the Palestinian problem. Through little effort on Arafat's part, the main rival identity—being Arab rather than Palestinian—was discredited. Arafat quickly tried to energize Palestinians in the newly occupied territories in the West Bank and Gaza, believing that the Algerian model could now be applied. He and others went into the West Bank in disguise to organize a local network to perform commando operations. In theory, the conditions were ripe for revolution. The Palestinians were now under occupation, and the cause had considerable support from outside powers. Indeed, Israeli intelligence reported in 1967 that the shock of defeat had quickly worn off and that anger was growing (Black and Morris, 1991, p. 239).

An Aborted Insurgency

Arafat, however, was not able to stir up the Palestinians on the West Bank and Gaza. At the time, they had little interest in resistance to Israel, and Israel quickly shut down what little opposition Fatah and other groups were able to foment. Compounding Fatah's problem of weak local enthusiasm, Palestinian fighters who arrived from outside the territories were often careless. They failed to compartment activities and in general had large and loose cells that hindered secrecy. One group that trained in Jordan and Syria even wore the same distinctive kit when they tried to infiltrate the West Bank. A deputy of Arafat later blamed the failure on "the efficiency of the Israeli secret services and the carelessness of our fighters."[6] The "people's war" there failed completely (Tessler, 1994, pp. 424–425; Rubin, 1994, pp. 16–18, 26; Kimmerling and Migdal, 1994, p. 226; Black and Morris, 1991, pp. 241, 255).

With its networks in the West Bank and Gaza stillborn, Fatah was forced to operate from neighboring states. These included Jordan, Syria, and Lebanon, with the balance varying according to the whims of the regime, its strength, and its attitude toward Palestinian guerrillas at that particular time.

Violence proved tremendously important to Fatah in its effort to rise above its rivals. Early raids in 1965 accomplished little, but they set Fatah apart from the moribund PLO. As one PLO leader put it, "We decided that the only way to keep the idea of real struggle alive was to struggle" (Kimmerling and Migdal, 1994, pp. 218, 227). Although most Fatah actions

---

[6] As quoted in Black and Morris (1991, p. 241).

were failures militarily, the group's violent resistance proved inspiring to many Palestinians and much of the Arab world, particularly in contrast to the collapse of the Arab armies in the 1967 war. The battle of Karamah in 1968, for example—a battle the Palestinians have portrayed as a victory—led to the deaths of 21 Israeli soldiers, even though Israel ultimately won and closed the Fatah camp. Ironically, the Israeli response proved tremendously important here. Although Fatah bungled its initial attacks, the publicity Israel gave to them demonstrated that Fatah was willing to fight.

Despite these identity successes, Fatah suffered, in part because it was poorly organized. At first, Arafat and other original leaders disdained traditional organizational structures. Thus, when Fatah attempted to act in the West Bank and Gaza, it lacked a local, grassroots infrastructure. This problem was worsened by the movement's presence in multiple countries, where the local leaders often operated independently from the central leadership, with full support of the host government. Even when the organization became much larger and more bureaucratized, power remained highly personalized, with Arafat often creating a confusing structure that overlapped, fostered corruption, and inhibited the coherence of the movement as a whole (Samuels, 2005; Kimmerling and Migdal, 1994, p. 216).

Organizational rivalry between Fatah and its rivals was tremendous and often debilitating. Initially, the PLO under Shuqayri was a tremendous challenge, and it took the debacle of the 1967 war to force a shift (Kimmerling and Migdal, 1994, p. 218). But even after Fatah became the leading organization within the PLO, it was not able to impose its will. Many Palestinian groups saw Arab regimes or Western targets as legitimate, and as a result, they attacked targets that Fatah did not endorse. This discredited the movement among many in the West and prevented it from acting in a coherent strategic way (Rubin, 1994, pp. 24–26). Nor did Fatah and other groups share intelligence or coordinate operations in any regular way (Black and Morris, 1991, p. 257). State sponsorship compounded this problem. Arab leaders repeatedly supported Arafat's rivals as a way of weakening the Palestinian leader and ensuring that their own objectives were met. As Abu Iyad, a key Palestinian leader, complained, the proliferation of small groups "practically strangled us" (Rubin, 1994, p. 36).

Relations with Outside Powers

Fatah's relations with the states that hosted Palestinians were contentious. The PLO made Jordan a major base until 1970. Radicals within the Palestinian movement (inspired by their revolutionary ideology that opposed all conservative Arab regimes, including that of Jordan's King Hussein) tried to push a confrontation with the Jordanian government, which was also inspired to act by the constant Israeli attacks on Palestinians operating from Jordanian territory. The Jordanian army cracked down and eventually drove the PLO to Lebanon. There the PLO had more success, as the Lebanese government was too weak to suppress it. Israel's cross-border raids, however, led the group to curtail its activities by the end of the 1970s. Israel's 1982 invasion proved a major blow to the PLO, killing many of its cadre and forcing others into exile. Even more important, it led Syria in particular to try to divide the movement and seize control, leading to a bloody clash within the Palestinian movement.

In general, the PLO relied heavily on Arab regimes for money and sanctuary, and at times for training. Yet this very reliance proved debilitating, as the states often limited PLO activities, prevented it from making useful concessions, or abandoned it at key times (Rubin, 1994, p. 128). Foreign states, however, did allow the PLO tremendous freedom of action on the creation of a Palestinian identity. In Kuwait, the PLO ran schools, enabling them to create Palestinian-only activities and teach Palestinian history. Similarly, in many refugee camps, the PLO was dominant and could push its interpretation of the appropriate identity and the proper cause. In the camps, the PLO was able to impose taxes, conscript men to fight, run the courts, establish unions, create women's organizations, and otherwise organize Palestinians under its banner (Kimmerling and Migdal, 1994, pp. 229–234; Tessler, 1994, p. 426).

The Israeli Response

Israel successfully shut down the PLO's attempted insurgency, but it was far less effective in offsetting the broader national movement. After the 1967 war, Israel's security services quickly established a network of informants. Israel punished villages that harbored Fatah members and steadily rolled up the new networks, killing 200 members and capturing 1,000 (Tessler, 1994, p. 425). Israel's ability to act outside its borders and the West Bank and Gaza was limited to strikes on suspected terrorist facilities. Going after schools, courts, and other important parts of identity- and cause-building was well beyond Israel's intentions or capabilities.

Israel faced an important tension between its day-to-day counterterrorism operations and its overall attempts to defeat the Palestinian movement. When Israel was present in Palestinian camps and otherwise occupying major population centers, its presence was highly intrusive and liable to anger Palestinians. When it departed these areas, however, it lost vital intelligence sources and made it far easier for the terrorist groups to operate freely.

Although the PLO failed to mount a successful insurgency, it was able to create a strong national movement, and this in the end led to a form of political success. This movement, in essence, put the Palestinian cause at the forefront of the world's agenda—a remarkable accomplishment given the relatively small numbers of people involved. Moreover, it proved consistently able to use international terrorism from its bases throughout the Arab world and in Europe. Thus it was able to force world attention to the Palestinian problem. After decades of struggle, following the signing of the Oslo Accords in 1993, Israel agreed to negotiate with the PLO and allow Arafat to return to the Palestinian territories as a legitimate leader.

The Failure of the Islamist Terrorists in Egypt

The experience of Egypt's Al-Jihad group (EIJ) and the larger and for many years more dangerous IG, stands in direct contrast to that of the Lebanese Hizballah and Fatah. EIJ and the IG also advocated a credo of Islamic revolution, but unlike Hizballah, they were unable to make the transition from terrorist group to insurgent movement.[7] And the shift of the struggle

[7] An important distinction between EIJ and the IG is in their attitudes toward popular revolution. EIJ's strategy was to have a small vanguard of militants that would mount a coup and, from the top, Islamicize society. The IG, in contrast, believed more in da'wa (proselytizing and education) as a way of creating a more Islamic society and from there creating an Islamic state (see International Crisis Group, 2004a, p. 6).

outside Egypt's borders led the movement to go more toward terrorism and further away from direct involvement in guerrilla war, unlike Fatah.

Religious militancy has recurred regularly in Egypt's long history, most recently in the 1970s. That period saw a revival both in popular Islam among the Egyptian people and in the proliferation of small Islamist groups with a revolutionary and violent agenda. After the assassination of Egyptian President Sadat in 1981, a harsh clampdown shattered many of these groups. However, others persisted in their opposition to the state. Confusingly, many groups revived the names of earlier groups, even though membership and goals often changed considerably over time.

In the early 1990s, the IG and EIJ appeared to be on the verge of creating a successful insurgency. Between 1990 and 1996, Egypt suffered its bloodiest period of strife in the 20th century, with over 1,200 dying. Terrorists targeted regime figures, secular intellectuals, foreign tourists, and Egypt's large Coptic Christian community (Ajami, 1998, pp. 200–204; Kepel, 2002, pp. 275–289). In upper Egypt, the IG operated openly and defied the state with its regular attacks (Kepel, 2002, p. 276).

More broadly, Islamism was doing well in Egypt, and the radical groups appeared to have developed a message in tune with most of the people. Many among Egypt's middle class embraced their faith: Businesses stopped for prayer, and women at universities began to go about veiled, in contrast to life in previous decades. Religious leaders who were once largely apolitical or puppets of the state now began to challenge the government, particularly on social issues (Abdo, 2000, pp. 4–6).

But by 1996, the tide had turned, and by 1998 the Egyptian radicals were reeling—the brutal 1997 attack that killed 58 tourists at the Luxor Temple was more a sign of the group's desperation than of a robust movement. By 1998, both groups had split into different factions, and many IG leaders called for a ceasefire in 1997.

Explaining the Militants' Failure

The use of violence by EIJ and the IG backfired. The violence alienated many Egyptians who supported an Islamist political agenda, with polls indicating a clear public distinction between groups that embraced violence and those that did not (Abdo, 2000, p. 14). The killing of innocents and the loss of tourism revenue proved particularly troubling to many Egyptians (Kepel, 2002, p. 289).

Egyptians also have strong national and Arab identities. Thus, while Islamist sentiment was on the rise, it competed with other established identities, many of which the state used to undercut the Islamists by branding them as tied to foreign governments such as Iran and Sudan.

Organizational rivalry was another problem. EIJ and the IG competed bitterly against one another for recruits and money. In addition, both competed with the Muslim Brotherhood and with the religious establishment, all of which claimed to be the true voice of Islam in Egypt.

State sponsorship of EIJ and the IG was always limited. Sudan hosted both groups in the early 1990s, and money came in from supporters in Saudi Arabia. Some important leaders, including the IG's spiritual guide Shaykh Omar abd-al Rahman (the "blind Shaykh"), operated from sanctuaries abroad—in Rahman's case, the United States. By the mid-1990s, this limited assistance was declining. Sudan steadily decreased its support for foreign radicals, particularly after a botched 1995 assassination attempt on Mubarak enabled Egypt and the United States to impose sanctions on Khartoum. Rahman's involvement in terrorism led to his imprisonment in the United States. The violence against civilians also alienated foreign backers, who sent their money elsewhere.

The Egyptian state exacerbated many of these identity, organizational, and foreign-support problems. At home, the state used a mix of co-optation and harsh repression to crush the extremists. The regime arrested thousands of suspects, and many were held without charge and treated brutally. The regime also tried to shut down mosques that did not have state support and to ban firebrand preachers from the pulpit. Even groups such as the Muslim Brotherhood, which opposed the regime but appeared to have reached a *modus vivendi* with it, suffered a major clampdown.

At the same time, the regime allowed other clerics, including those with a strong anti-Western agenda, access to the state-controlled television. Religious figures began to censor books and movies as well (Abdo, 2000, pp. 14–22, 78; International Crisis Group, 2004a, p. 13). Over time, the regime reached out to the Muslim Brotherhood, allowing it a significant presence in the political system, as well as more freedom to organize (International Crisis Group, 2004a, p. 1). The Islamist identity thus had an outlet, which reduced its salience as a grievance.

The lack of a foreign haven, domestic sanctuary, or other means to protect the movement's leadership proved devastating. The EIJ and IG mistakes and the regime's ruthlessness meant that a botched operation often led to the arrests of dozens of operatives. These arrests, in turn, spiraled and enabled the regime to wrap up hundreds of activists. Over time, the experienced cadre were dead, in jail, or in exile. The new terrorist leaders and operatives repeatedly made mistakes, making the group as a whole far less effective and far more vulnerable to counterterrorism.

In contrast to Fatah and Hizballah, the Egyptian groups faced a huge disadvantage in that they were not able to harness nationalism effectively. Fatah and Hizballah both directed much of their military effort, and even more of their propaganda, against Israel (and, in Hizballah's case, the United States). Israel occupied historic Palestine and parts of Lebanon, making it easy to beat the drum of nationalism. The Egyptian groups were fighting against their own government, which is a far more difficult situation for groups trying to stir up latent nationalism.

The Costs of Success

As the rebellion petered out in Egypt, the international nature of the groups' agendas grew. In 1995, Egyptian radicals operating from Sudan attempted to assassinate Egyptian President Mubarak on a trip to Addis Ababa. Later that year, they bombed the Egyptian embassy in Islamabad, killing 16 people. This shift to Egyptian targets abroad soon led to a focus on non-Egyptian, particularly American, targets. Zawahiri led much of the EIJ membership to join al-Qa'ida, largely because of the difficulty of conducting operations in Egypt. By 2001, much of al-Qa'ida's most important cadre had come from EIJ.

A second cost was the impact on Egyptian society. By defeating the radicals through co-optation as well as coercion, the secular nature of Egyptian society declined. Secular intellectuals and secular political parties both suffered tremendously. In addition, the regime's willingness to bend laws in the name of antiterrorism undermined the already weak legitimacy of Egypt's judiciary.

A third cost was that terrorism continued even as the insurgency faded. The Luxor attack was the bloodiest but by no means the last. In 2005 and 2006, jihadist groups, probably affiliated with or at least inspired by al-Qa'ida, conducted a number of attacks on tourist and other facilities on the Sinai peninsula.

APPENDIX B

Applying the Proto-Insurgency Concept to Saudi Arabia Today

Since May 2003, when Saudi radicals linked to al-Qa'ida conducted attacks on compounds housing U.S. security personnel in the Kingdom, killing 34 people, the Kingdom has been in a state of low-level civil strife that some analysts have called an insurgency. Saudi terrorists, often called Al-Qa'ida of the Arabian Peninsula (QAP), have attacked U.S. facilities, foreign workers in the Kingdom, and regime security forces, among other targets. Although data on any developments in the Kingdom are scarce, media reporting and regime statements indicate that more than 200 troops and militants have died in firefights and attacks and the Saudi police and security services have arrested hundreds more.[1] The terrorists appear to see bin Laden as a guide, but there does not appear to be a direct operational role for the al-Qa'ida leadership.

Characterizing the strife as an insurgency, however, overlooks two important dimensions of that term: size and the use of guerrilla war. Like many insurgent movements, QAP uses terrorism and tries to mobilize people against the regime. However, its size is at the borderline between a terrorist movement and an insurgency. Based on casualty and arrest figures, it seems that QAP has several hundred active members—comparable to the size of the PIRA or the Red Brigades. Such numbers are large, but most insurgencies are much larger. In Iraq, for example, the insurgency has thousands of fighters under arms. Even more important, QAP has not been able to conduct guerrilla warfare. On balance, this suggests that QAP is more of a proto-insurgency. If it were a little larger and could conduct guerrilla warfare, it would be a true insurgency, but its already considerable size and use of political mobilization and sustained terrorism suggest that it is more than a typical terrorist group.

Mixed Progress on Building a Cause

QAP and other jihadist groups promulgate the idea that political loyalty should be to an Islamic ideal rather than to the Saudi nation-state. They believe that Saudi Arabia should be governed according to Islamic law and that loyalty should be given to jihadist religious and political leaders, not to the Al Saud. They embrace *salafi* ideas that emphasize the rejection of theological innovation and the need to emulate the early leaders of Islam. QAP, like many al-Qa'ida affiliates, tries to square the circle of a transnational ideology and local nationalism by

[1] For a review, see Cordesman and Obaid (2005).

limiting the scope of its activities to one country—in this case, Saudi Arabia—even though its ideology is much broader.

Ironically, the identity QAP and its allies champion is quite similar to that of the Saudi regime itself, which has somewhat limited the group's appeal. The regime has long styled itself as a government that governs according to Islamic law and that has a strong religious identity. Indeed, Saudi radicals and the Al Saud both invoke the same theologians and teachers to justify their actions. QAP rejects the religious bona fides of the Al Saud, but creating an identity based on Islam rather than secular nationalism has already been done.

This argument appeared to gain support in the early 1990s, when clerics associated with *al-sahwa al-Islamiyya* (the Islamic awakening) blasted the Saudi regime and the official clergy, particularly about the regime's ties to the United States and the presence of U.S. forces on the Arabian peninsula. The arrest of the anti-regime clerics silenced this criticism at home, but many radicals viewed it as proof that the Al Saud opposed true Islam. Bin Laden himself praised several of these clerics and noted that the regime's arrest of them was further proof of its perfidy.[2]

Many Saudis consider themselves to have an Islamic identity rather than a national one, and some of these have embraced the jihadists. As is endlessly noted, 15 of the 19 hijackers on 9/11 were Saudis. Many estimates of al-Qa'ida's composition place Saudis at or near the top in terms of numbers within the organization. Literally thousands of Saudis trained in Afghanistan, and they are a deep bench for the jihadists to draw upon. The regime's regular arrests of dozens of suspects suggests the large number of potential fighters—even massive arrests have not crippled the movement. Moreover, the 9/11 Commission found that the Kingdom was al-Qa'ida's greatest source of funding (National Commission on Terrorist Attacks Upon the United States, n.d.).

Perhaps the would-be insurgents' biggest problem is that they have not captured the allegiance of most *salafis* in the Kingdom, let alone most Saudis—a key barrier to their future expansion. The violence of the jihadists in November 2003 in particular, which involved attacks on Muslim and Arab targets and subsequent attacks on Saudi policemen and soldiers, led to widespread condemnation. Even former *sahwa* firebrands such as Safar al-Hawali and Salman al-'Awda—shaykhs whom bin Laden himself had praised in the early 1990s—condemned the May attacks (Dekmejian, 2003). QAP has since moved away from controversial attacks on Arab and Muslim targets, but much of the damage to its image appears to have been done.

Perhaps an even larger problem is that many *salafis* in Saudi Arabia embrace the identity of being a Muslim rather than a Saudi but do not accept this as a political cause. Far more Saudi "radicals" appear to be quietist in their approach. They reject the Saudi regime, but they also reject QAP and other jihadists. They see jihadists as far too political and thus inherently corrupt. In general, they believe that political organization is dangerous, because it leads individuals to place the temporal over the spiritual.

---

[2] An excellent review of divisions within the religious communities in Saudi Arabia can be found in International Crisis Group (2005).

The Saudi State Response

The Saudi regime has effectively fought the proto-insurgency, hindering its effort to spread its grievances to the population at large. Saudi security forces have engaged in widespread arrests and have decimated the senior ranks of the jihadists in the Arabian peninsula. Since the 2003 attacks, the regime has regularly issued "most wanted" lists of militants and has methodically killed or arrested those on the lists. The jihadist attacks have also prompted the regime to take the problem the jihadists pose worldwide much more seriously.

After the violence in 2003, the Saudis implemented a number of unprecedented measures to fight terrorism. The regime excised much, though not all, of the material denigrating other religions from school textbooks. A senior U.S. Treasury Department official noted that the Saudis increased their regulation of informal money transfers, stepped up fund-management responsibility, and increased prohibitions on charitable donations outside the Kingdom. The regime publicized a list of names and photos of the most-wanted terrorist suspects and visibly increased security—very public measures for a regime that prefers to operate in the background. Then–Crown Prince Abdullah also traveled to Russia and condemned the Chechens' violence. These measures suggest that the Al Saud recognized the connections among disparate Islamists, even those not directly attacking the Kingdom, and the way their proselytizing bolsters al-Qa'ida. Testifying in March 2004, Ambassador Cofer Black, then the U.S. Coordinator for Counterterrorism, declared that the Saudis understood the threat they faced and were closely cooperating with U.S. officials (Black, 2004).

The Saudi regime has effectively engaged in a "war of ideas." By converting several prominent clerics from adversaries and critics to regime supporters, it was able to use highly credible voices to undercut support for the radicals. The jihadists' own missteps in targeting, such as killing Muslims and Arabs in early attacks, made it easier for the regime to gain the backing of its former critics and to paint the jihadists as murderous thugs in the eyes of the Saudi people. The Saudi government also lined up clergy more openly sympathetic to the regime to condemn the jihadists and pressed hard to ensure that they offered a united front against them. It also pushed Hamas and other radical organizations it backs to condemn the violence. Finally, the regime published confessions of captured terrorists to show their ignorance and brutality.

In terms of identity, the regime has tried to portray itself rather than the militants as the true standard bearers of Islam. In part, this has involved garnering statements of support from various religious leaders and continuing to support religious activities both in the Kingdom and outside it. The regime has also attacked the behavior of the jihadists as un-Islamic.

Regime skill and capacity remain a constant concern. Corruption is rampant in Saudi Arabia, and the quality of governance is poor (Transparency International, 2004). The Saudi military is often inefficient, particularly in the use of technology and coordination of units. Moreover, al-Qa'ida has made a conscious effort to cultivate military and government officials in Saudi Arabia, an effort that its local affiliate has also presumably made, suggesting that QAP penetration is a serious problem (Anonymous, 2003, p. 22).

The overall Saudi campaign is further hindered by the fact that the regime's legitimacy, while not weak, is under strain. Soaring oil prices have helped reduce economic concerns, but structural problems remain deep, and little progress has been made. Leadership is also a long-term concern. King Abdullah is respected, but he is old, and his potential successors are less impressive: Sultan is venal and viewed as incompetent, while Nayif is venal and brutal. These potential successors, too, are old, and the Kingdom could experience a rapid set of leadership changes in the years to come that could lead to political stagnation or paralysis as new leaders try to consolidate their positions before acting.

Limited Outside Aid . . . For Now

The terrorists lack a haven inside the Kingdom at present. The desert terrain is not well-suited for hiding, and they do not have enough support in urban areas to deny the government entry. As a result, the regime has been able to arrest both leaders and followers on a steady basis.

Externally, no state supports QAP.[3] However, QAP and like-minded groups are likely to draw support from what remains of bin Laden's networks and from new ones that have developed in recent years. Bin Laden has always closely followed events in the Kingdom, and many analysts have suggested that the 9/11 attacks and other operations were directed in part to influence events there. Because of the Kingdom's importance as the home of Islam, Muslim radicals view events there as vital to the overall cause.

Iraq poses a tremendous challenge to the Saudi government's effort, for several reasons. The continued U.S. military presence there serves as a recruiting device for the broader jihadist movement. In part because of the anger the U.S. presence generates, the overall pool of radicals is wide in Saudi Arabia. The close Saudi ties to the United States worsen this problem. Potential allies of the Al Saud distance themselves from the regime on this issue, which "proves" to many Islamists that the regime itself is corrupt and practices what its critics have long called "American Islam." Other effects are more immediate. The chaos in Iraq has enabled jihadists there to carve out parts of the country as a mini-haven where they are teaching new improvised explosive device (IED) techniques, inculcating young recruits into their global view of the enemy, forming new networks, and otherwise advancing their cause. Saudis who go to Iraq will return as more formidable foes.

In addition, Iraq may provide the haven QAP needs to turn its terrorist movement into a full-blown insurgency. The hotel attacks in Jordan, which were orchestrated from Iraq, may be a harbinger of similar attacks in the Kingdom. The open border and long-standing ties between Saudi and Iraqi Sunni tribes make it particularly easy for jihadists to travel between the two countries.

The Complicated U.S. Role

U.S. intelligence agencies are currently working with their Saudi counterparts to strengthen regime capacity in gathering, processing, and using intelligence and in tracking terrorist

[3] The May 2003 attacks appear to have been conducted with the assistance of al-Qa'ida figures operating from Tehran. The degree of Iranian complicity in these attacks remains unclear, but Iran has not been implicated in the subsequent campaign of violence.

financing. Such measures should continue. The United States should also try to train the Saudi National Guard and police when requested. Whenever possible, this training should occur outside the Kingdom, through third parties, or by other means that will minimize the visibility of the U.S. role given the strong anti-U.S. sentiment in the Kingdom.

Far less important is U.S. military assistance to the Kingdom. Currently, the Kingdom faces at best a weak conventional military threat from Iran and no threat from Iraq, its traditional security concern. Spending money on expensive systems and engaging in politically unpopular military cooperation with the United States may thus be counterproductive for the overall security of the Kingdom.

The United States will have a major role in limiting the fallout from Iraq, particularly if the strife increases. Several hundred fighters based in Iraq that target the Saudi regime could dramatically weaken the stability of the Kingdom. Helping improve border security and limiting refugee flows are key missions.

Whenever possible, U.S.-Saudi cooperation must remain in the background. Polls in the Kingdom regularly show U.S. policy to be widely loathed, with less than 20 percent (usually far less) of Saudis having a favorable view of the United States. Because insurgents could capitalize on cooperation with the United States to undercut the regime on nationalistic grounds, Washington must work behind the scenes.

One way for the regime to improve its standing at home would be to distance itself further from U.S. policy. In the past, the regime has often tried to defeat its opponents by co-opting their cause. If violence increases, the Al Saud may decide to take the wind out of the opposition's sails by becoming more vocal in its opposition to U.S. policies and by cutting cooperation. Should the insurgency in Iraq worsen, the United States must prepare for a decline in Saudi Arabia's public cooperation with Washington and an increase in criticism. Al Saud leaders will still recognize the importance of the United States, but on high-visibility issues that are not vital to the Kingdom's security, they may decide to openly reject it.

APPENDIX C

Proto-Insurgency Indicators

Indicators for proto-insurgencies fall into two general categories: indicators that a proto-insurgency may break out and indicators that one may become a full-blown insurgent movement.[1] However, because proto-insurgencies involve small numbers of people, many of whom are divorced from the population as a whole, societywide measures generally provide few benefits for judging when a proto-insurgency will arise. Some possible indicators of developing proto-insurgencies are briefly noted at the end of this appendix, but the focus here is on *measures that indicate when proto-insurgencies may grow into full-blown insurgent movements*. Thus, the measures assume that at least a few individuals already exist who have turned to violence and meet the definition of a proto-insurgency but their group has not turned into a large and capable insurgent movement.

Indicators to consider when trying to determine whether a proto-insurgency is able to make the transition to a full-blown insurgency include (1) the strength of the proffered identity, (2) group composition, (3) relations with other community members, (4) use of and response to violence, (5) existence of a sanctuary, (6) external support, and (7) the state response.[2]

Identity Indicators

Proto-insurgencies are engaged in a battle of identities. They are trying both to promote their preferred identity (Islamist, ethnic, etc.) and to undermine rival identities proposed by the state or other community members. Several indicators shed light on how well a proto-insurgent is faring in this battle:

- The strength of the state identity. Does the population consider itself first and foremost to be members of the nation championed by the state? How strong is patriotic sentiment? Are national holidays enthusiastically celebrated? Is the dominant literature of the country in the language proposed by the state? Are there rival cultural elites that do not accept

[1] Another RAND project is examining correlates for insurgent success and will provide indicators for the development and success of insurgencies in general.

[2] The CIA *Guide to the Analysis of Insurgency* (C.I.A., n.d., pp. 6–10) examines indicators for incipient insurgency, which "encompasses the pre-insurgency and organizational stages of an insurgency conflict." A number of the measures discussed in this appendix are taken from this source.

the national identity (and are there national cultural elites that disparage other identities)? Do all members of society believe they have a shared history?

- The strength of alternative identities. Using similar measures, what are the strengths of rival identities, such as tribe, ethnicity, and religion?
- Whether attacks (criminal, political, etc.) on one member of the population provoke outrage from individuals of different tribes, religious communities, or ethnic groups who do not know the victim.
- How identity measures have changed over time. Which ones are getting stronger and which ones are getting weaker? Under what conditions do some identities fluctuate? Is there important regional variation?
- Whether the would-be insurgent group is able to harness nationalism or a similar "us-versus-an-outsider" dynamic in its contest. Is the state perceived as being led by people who are resistant to foreign pressure, or are state leaders perceived as too close to Washington or another foreign power?
- Whether the media exalt individuals and causes that the group claims as its own (e.g., Saudi media glorifying the exploits of the Mujahedin in Afghanistan and the Balkans).
- Whether the group controls generators of identity. Who controls schooling? What are the language policies in the country? Are there movies and books in the language the proto-insurgents champion?

Group-Composition Indicators

Although considerable attention focuses on the cause championed by proto-insurgents, success or failure often comes down to a question of group dynamics: Can the group recruit effectively, and does it have the resources to wage a full-blown insurgency?

- Size. One of the definitional distinctions between a proto-insurgency and a full-blown insurgency is the size of the group. Going beyond an active membership in the dozens to a membership in the hundreds is one clear indicator. It is always necessary to track the numbers of full-time cadre, part-time cadre, active supporters, and potential supporters.
- Membership motivation. Members of small terrorist or insurrectionary groups are often highly idealistic—a necessary characteristic given the overwhelming odds in favor of the state and the high level of danger the individuals face. Large-scale insurgencies, however, have many members who joined because they were coerced or because they saw a chance of personal gain from membership. The more members who joined because of fear or reward, as opposed to idealism, the more likely the insurgency is in the process of becoming full-blown.
- Breadth of membership. Many small groups are bound by tribe, family, and region or are otherwise limited in who becomes a member. Such networks play a vital role early in an insurgency in helping ensure operational security and decreasing barriers to trust. Success, however, requires transcending this narrow base. If a group is able to draw on multiple subidentities, it is more likely to develop into a full-blown insurgency.

- Weapons and materiel. How large are the weapons and materiel caches of the group? Is the group able to replace lost weapons? How sophisticated are the weapons?
- Whether the group is able to "tax" parts of the population. If so, how much of this taxation is voluntary? How much is involuntary? (Both are useful, as the insurgent must be able to use both suasion and fear, but if individuals are giving *despite* their preferences, this suggests a high degree of insurgent strength.) In which regions is the group able to collect taxes?
- How much money the group has access to. Is it able to pay followers? Can it pay followers more than other likely economic opportunities can? Is it able to bribe officials?

Indicators of Relations with Other Community Members

What is the relationship between the group in question and the broader movement as a whole? The prevalence of smaller, more-radical groups paralyzed Fatah, hindering its ability to marshal all the resources available to the Palestinian nationalist cause. Similarly, many small groups have been unable to develop into broader movements because peaceful political movements were able to effectively advance their agendas. Community indicators include:

- The popularity of the overall cause the group espouses. How many people show up at demonstrations? How many people are members of nonviolent political and social organizations with a similar agenda?
- Relations with rivals. Is the group incorporating other small rivals (as Hizballah did), or is it constantly dividing into smaller groups (as happened to many Pakistan-backed groups fighting in Kashmir)?
- The attitudes of key cultural and religious figures who are not part of the group. Do they endorse the group's activities and see it as legitimate? Does the group have an informal presence in unions, political parties, churches/mosques, or other legitimate social organizations?
- How tenable the moderate option is. Is peaceful political change plausible? Likely?
- Whether the group is able to coordinate its activities with a broader political movement. Is the group able to dominate that movement to the point that the broader movement acts only with the group's tacit approval?
- Whether members of the moderate movement are joining the radical wing or vice-versa. Are leaders moving from one camp to another?
- How members of a cause or movement who do not embrace violence see the group. Are the group's activities beyond the pale, or are the members seen as Robin Hoods?
- The extent of group propaganda and proselytization efforts. Are they expanding to new areas?
- Whether the group is able to establish a social services network to extend its reach and popularity.

Indicators of the Use of and Response to Violence

The prevalence of violence is an indicator, albeit an imperfect one, of the relative strength of a proto-insurgency. Similarly, the most skilled groups are able to plan for the state response to violence and use it to advance their cause. The following are indicators of a group's use of violence and ability to deal with the state response:

- The rate of attacks on government forces and on the civilian population. Is the group able to attack guarded or other "hard" targets?
- The extent of the group's area of operations.
- Whether armed fighters are able to show themselves openly. In how wide an area?
- Violence against diplomats or other key figures abroad.
- The discipline of the fighters. Do they conserve ammunition, recover bodies of fallen comrades, and otherwise display a degree of professionalism?

After a group uses violence, the reactions of the population often indicate the strength and potential influence of the group:

- Do civilians flee after attacks? If the state is trusted and the group is perceived to be weak, the use of violence will not shake civilian confidence in police and security services. Thus, the populace is more likely to stay in areas despite the risk of violence.
- How are failed attacks perceived? When potential support is deep, even failed attacks can be successfully portrayed as heroic attempts to resist the government. When the group's attack does not succeed tactically, is it able to convert it into a strategic success?
- How does the group respond to initial arrests of its members? Has it planned for their replacement? Are its security procedures robust enough to ensure that low-level arrests do not devastate the group? What is the impact on group morale?
- Does the population at large support increased control and repression measures after a violent attack? Does this support extend to the communities the would-be insurgents seek to attract?
- Is the group able to sustain its campaign of violence? Is it able to extend its reach into new areas?

Indicators of Sanctuary

In order to flourish, groups often need access to a sanctuary, a place to hide in and grow in the face of government countermeasures. (Indicators of external sponsors providing sanctuary are presented in the next section.) Indicators of sanctuary include:

- Whether the terrain is suitable for guerrilla warfare. What is the overall presence of mountainous, jungle, or other less accessible forms of terrain? Are these areas near where the initial cadre of the group is based? Are they easily cordoned off?

- Whether there are refugee camps outside the control of the government that the group can exploit.
- Whether parts of the country constitute a "no-go" zone for police and security forces due to violence, ethnic antipathy, or insurgent activity. What is the size of this zone and how is it changing?
- Whether the insurgents are able to sleep or rest in towns and villages outside the sanctuary area.

Indicators of External-State-Support Effectiveness

External support can be both a blessing and a curse: It can greatly boost a group's capabilities and provide an invaluable sanctuary, but at the same time it can reduce the group's freedom of movement and overall appeal. Indicators of state support include the following:

- The type and scale of support provided. Is the group receiving assistance in operational security? Operational planning? Logistics? Financial support? Do large numbers of group members travel to receive such assistance?
- Whether the group can draw on diaspora support. How much support is provided and what is the overall potential for more?
- Cooperation with group members abroad. Do group members in different foreign sanctuaries work together well, or do the different powers try to use them as proxies against other external backers?
- The extent of constraints imposed by the external sponsor. Do the constraints limit the type and nature of the group's attacks or other activities for reasons that are tied to the sponsor's concerns (as opposed to helping the group make better decisions)?
- Whether the sponsor seeks to control the overall movement. Does the state sponsor divide the movement into smaller groups in order to better assure its own control?
- Whether the foreign sponsorship decreases the group's legitimacy among different segments of the population and among potential core supporters.
- Whether state support is a substitute for local strength. Does the group have local networks that the foreign support augments?

State-Response Indicators

Perhaps the greatest source of growth for a proto-insurgent group is a clumsy state response to violence, in contrast to a deft crackdown that can end a group once and for all. Measures to consider include:

- The government's flexibility regarding the identity and grievance the group seeks to exploit. Does the government recognize the need to meet some of the grievances being advanced? Is the government able to co-opt elements of the group's cause? Is the government able to harness nationalism in its response?

- The skill of the police and intelligence services. Do these services understand the need to use force in a restrained way and in combination with precise intelligence? Do the intelligence services have information on all segments of society?
- The effectiveness of the administration and the bureaucracy. Can they deliver services? Can they collect taxes?
- The overall level of corruption in the country. Do the police and security services reflect the overall level of corruption in society?
- Popular faith in the bureaucracy and the police. Do the police have the trust of local communities?
- Whether the government is able to exploit, and perhaps create, divisions within the opposing movement as a whole.
- Whether the government is willing and able to distinguish between peaceful opponents and violent ones. Do its policies allow moderate politicians to flourish?
- Whether the proto-insurgency has successfully penetrated parts of the government in the region where it is most active. Has it done so in other regions? Has it infiltrated the police and intelligence services?

Bibliography

Abdo, Geneive, *No God but God: Egypt and the Triumph of Islam*, New York: Oxford University Press, 2000.

Ajami, Fouad, *The Vanished Imam*, Ithaca, NY: Cornell University Press, 1986.

———, *The Dream Palace of the Arabs*, New York: Pantheon Books, 1998.

Anderson, Benedict, *Imagined Communities: Reflections on the Origins and Spread of Nationalism*, London: Verso, 1983.

Anonymous, *Through Our Enemies' Eyes: Osama Bin Laden, Radical Islam and the Future of America*, Washington, DC: Brassey's Inc., 2003.

Beckett, Ian F. W., *Modern Insurgencies and Counter-Insurgencies: Guerrillas and Their Opponents Since 1750*, New York: Routledge, 2003.

Bell, J. Bowyer, "The Armed Struggle and Underground Intelligence: An Overview," *Studies in Conflict and Terrorism*, Vol. 17, 1994, pp. 115–150.

———, *The Secret Army: The IRA*, New Brunswick, NJ: Transaction Publishers, 2003.

Black, Ian, and Benny Morris, *Israel's Secret Wars*, New York: Grove Press, 1991.

Black, J. Cofer, Coordinator for Counterterrorism, Testimony before the House Committee on International Relations, Subcommittee on the Middle East and Central Asia, March 24, 2004. As of January 25, 2007: http://www.state.gov/s/ct/rls/rm/2004/30740.htm.

Blanford, Nicholas, "Hizbullah Attacks Force Israel to Take a Hard Look," *Jane's Intelligence Review*, Vol. 11, No. 4, 1999. As of August 28, 2006:
http://jir.janes.com/public/jir/index.shtml.

Brynen, Don, *Sanctuary and Survival: The PLO in Lebanon*, Boulder, CO: Westview, 1990.

Buchta, Wilfreid, *Who Rules Iran*, Washington, DC: The Washington Institute for Near East Policy, 2001.

Byman, Daniel, "The Logic of Ethnic Terrorism," *Studies in Conflict and Terrorism*, Vol. 21, No. 2, Spring 1998, pp. 149–169.

———, *Keeping the Peace: Lasting Solutions to Ethnic Conflict*, Baltimore, MD: The Johns Hopkins University Press, 2002.

———, *Deadly Connections: States That Sponsor Terrorism*, New York: Cambridge University Press, 2006.

———, "Israel's Counterterrorism Operations Against Hezbollah," in Robert Art (ed.), *Comparative Counterterrorism Strategies*, Washington, DC: U.S. Institute of Peace, forthcoming.

Carnegie Commission on Preventing Deadly Conflict, *Preventing Deadly Conflict*, Washington, DC: Carnegie Commission on Preventing Deadly Conflict, 1997.

Central Intelligence Agency, *Guide to the Analysis of Insurgency*, Washington, DC (n.d.).

Collier, Paul, "Rebellion as a Quasi-Criminal Activity," *Journal of Conflict Resolution*, Vol. 44, 2000, pp. 839–853.

Collier, Paul, and Anke Hoeffler, "Greed and Grievance in Civil War," *Oxford Economic Papers*, Vol. 56, 2004.

Collier, Paul, Anke Hoeffler, and Nicholas Sambanis, "The Collier-Hoeffler Model of Civil War Onset and the Case Study Project Research Design," in Paul Collier and Nicholas Sambanis (eds.), *Understanding Civil War Volume II: Europe, Central Asia, and Other Regions*, Washington, DC: The World Bank, 2005.

Connor, Walker, *Ethnonationalism: The Quest for Understanding*, Princeton, NJ: Princeton University Press, 1994.

Cordesman, Anthony H., and Nawaf Obaid, *Al-Qaeda in Saudi Arabia*, Washington, DC: Center for Strategic and International Studies, January 26, 2005. As of August 23, 2006:
http://csis.org.

Crenshaw, Martha, "An Organizational Approach to the Analysis of Political Terrorism," *Orbis,* Vol. 29, No. 3, Fall 1985, pp. 473–487.

——— (ed.), *Terrorism in Context*, University Park, PA: Pennsylvania State University Press, 1995.

Dekmejian, Richard, "The Liberal Impulse in Saudi Arabia," *The Middle East Journal,* Vol. 57, No. 3, Summer 2003, pp. 400–413.

Della Porta, Donatella, "Left Wing Terrorism in Italy," in Martha Crenshaw (ed.), *Terrorism in Context*, University Park, PA: Pennsylvania State University Press, 1995, pp. 105–159.

Eshel, David, "Counterguerrilla Warfare in South Lebanon," *Marine Corps Gazette,* Vol. 81, No. 7, July 1997, pp. 40–45.

Fearon, James, and David D. Laitin, "Explaining Interethnic Cooperation," *American Political Science Review,* Vol. 90, No. 4, 1996, pp. 715–735.

———, "Ethnicity, Insurgency, and Civil War," *American Political Science Review,* Vol. 97, No. 1, February 2003, pp. 75–90.

Fuller, Graham, *The Future of Political Islam,* New York: Palgrave MacMillan, 2004.

Galula, David, *Counterinsurgency Warfare: Theory and Practice*, New York: Praeger, 1964 (reprinted in 2005).

Greenfeld, Liah, *Nationalism: Five Roads to Modernity*, Cambridge, MA: Harvard University Press, 2003.

Hajjar, Sami G., "Hizballah: Terrorism, National Liberation, or Menace?" *Strategic Studies Institute*, August 2002. As of August 23, 2006:
http://www.strategicstudiesinstitute.army.mil/pdffiles/PUB184.pdf.

Hall, Raymond (ed.), *Ethnic Autonomy: Comparative Dynamics*, New York: Pergamon Press, 1979.

Hamzeh, A. Nizar, "Lebanon's Hizbullah: From Islamic Revolution to Parliamentary Accommodation," *Third World Quarterly,* Vol. 14, No. 2, 1993, pp. 321–337.

———, "Islamism in Lebanon: A Guide," *Middle East Review of International Affairs,* Vol. 1, No. 3, Spring 1997. As of August 28, 2006:
http://meria.biu.ac.il/journal/1997/issue3/jv1n3a2.html.

Harik, Judith Palmer, "Between Islam and the System: Sources and Implications of Popular Support for Lebanon's Hizballah," *The Journal of Conflict Resolution,* Vol. 40, No. 1, March 1996, pp. 41–67.

Higgins, Andrew, and Alan Cullison, "Terrorist's Odyssey," *The Wall Street Journal,* July 2, 2002, p. A1.

Hiro, Dilip, *Lebanon: Fire and Embers*, New York: St. Martin's Press, 1992.

Hoffman, Bruce, "The Modern Terrorist Mindset," in Russell D. Howard and Reid L. Sawyer (eds.), *Terrorism and Counterterrorism: Understanding the New Security Environment*, Gilford, CT: McGraw Hill, 2002a.

———, "Rethinking Terrorism and Counterterrorism Since 9/11," *Studies in Conflict & Terrorism,* Vol. 25, No. 5, 2002b, pp. 303–316.

———, "Insurgency and Counterinsurgency in Iraq," Santa Monica, CA: RAND Corporation, OP-127-IPC/CMEPP, 2004. As of January 18, 2007:
http://www.rand.org/pubs/occasional_papers/OP127/.

———, *Inside Terrorism*, New York: Columbia University Press, 2006.

Howard, Russell D., and Reid L. Sawyer (eds.), *Terrorism and Counterterrorism: Understanding the New Security Environment*, Guilford, CT: McGraw-Hill, 2002.

Hudson, Michael, "The Breakdown of Democracy in Lebanon," *Journal of International Affairs,* Vol. 38, No. 2, Winter 1985, pp. 277–282.

Human Rights Watch, *Civilian Pawns: Laws of War Violations and the Use of Weapons on the Israel-Lebanon Border*, New York: Human Rights Watch, 1996.

Husayn, Fu'ad, *Al-Quds Al-'Arabi*, June 8, 2005, p. 4.

International Crisis Group, Middle East Briefing No. 7, "Hizbollah: Rebel Without a Cause?" (30 July 2003). As of August 23, 2006:
http://www.crisisgroup.org/home/index.cfm?id=1828&l=1.

———, Middle East/North Africa Briefing No. 13, "Islamism in North Africa II: Egypt's Opportunity" (20 April 2004a). As of August 23, 2006:
http://www.crisisgroup.org/home/index.cfm?id=3713&l=1.

———, Asia Report No. 83, "Indonesia Backgrounder: Why Salafism and Terrorism Mostly Don't Mix" (13 September 2004b). As of August 23, 2006:
http://www.crisisgroup.org/home/index.cfm?id=2967&l=1.

———, Middle East Report No. 31, "Saudi Arabia Backgrounder: Who Are the Islamists?" (21 September 2005). As of August 23, 2006:
http://www.crisisgroup.org/home/index.cfm?id=3021&l=1.

———, Middle East Report No. 49, "Enter Hamas: The Challenge of Political Integration" (18 January 2006a). As of August 23, 2006:
http://www.crisisgroup.org/home/index.cfm?id=3886&l=1.

———, Middle East Report No. 52, "The Next Iraqi Civil War? Sectarianism and Civil Conflict" (27 February 2006b). As of August 23, 2006:
http://www.crisisgroup.org/home/index.cfm?l=1&id=3980.

Jaber, Hala, *Hezbollah: Born with a Vengeance*, New York: Columbia University Press, 1997.

Kepel, Gilles, *Jihad: The Trail of Political Islam*, Cambridge, MA: Belknap Press, 2002.

Kimmerling, Baruch, and Joel S. Migdal, *Palestinians: Making of a People*, Cambridge, MA: Harvard University Press, 1994.

Leites, Nathan, and Charles Wolf, Jr., *Rebellion and Authority: An Analytic Essay on Insurgent Conflicts*, Chicago, IL: Markham, 1970.

Lustick, Ian, "Stability in Deeply Divided Societies: Consociationalism Versus Control," *World Politics,* Vol. 31, No. 3, April 1979, pp. 325–344.

———, *Arabs in the Jewish State: Israel's Control of a National Minority*, Austin, TX: University of Texas Press, 1980.

McCord, Arline, and William McCord, "Ethnic Autonomy: A Socio-Historical Synthesis," in Raymond Hall (ed.), *Ethnic Autonomy: Comparative Dynamics*, New York: Pergamon Press, 1979.

McCormick, Gordon H., "Terrorist Decision Making," *Annual Review of Political Science,* Vol. 6, 2003, pp. 473–507.

McCormick, Gordon H., and Guillermo Owen, "Security and Coordination in a Clandestine Organization," *Mathematical and Computer Modeling,* Vol. 31, No. 6, March 2000, pp. 175–192.

National Commission on Terrorist Attacks Upon the United States, "Monograph on Terrorist Financing." As of August 23, 2006:
http://www.9-11pdp.org/.

Norton, Augustus Richard, *Amal and the Shi'a,* Austin, TX: University of Texas Press, 1987.

———, "Hizballah and the Israeli Withdrawal from Southern Lebanon," *Journal of Palestine Studies,* Vol. 30, No. 1, Autumn 2000. As of August 28, 2006:
www.jstor.org.

Pillar, Paul, *Terrorism and U.S. Foreign Policy,* Washington, DC: The Brookings Institution, 2001.

Posen, Barry, "Military Responses to Refugee Disasters," *International Security,* Vol. 218, No. 1, Summer 1996, pp. 72–111.

Ranstorp, Magnus, *Hizb 'allah in Lebanon: The Politics of the Western Hostage Crisis,* New York: St. Martin's Press, 1997.

Ricks, Thomas, *Fiasco: The American Military Adventure in Iraq,* New York: Penguin Press, 2006.

Rosenau, William, "The Kennedy Administration, U.S. Foreign Internal Security Assistance and the Challenge of 'Subterranean War,'" *Small Wars and Insurgencies,* Vol. 14, No. 3, Autumn 2003.

Rubin, Barry, *Revolution Until Victory? The Politics and History of the PLO,* Cambridge, MA: Harvard University Press, 1994.

Samraoui, Mohammed, *Chronique des Années de Sang: Algérie: Comment les Services Secret Ont Manipulé les Groupes Islamistes,* Denoël Impacts, 2003.

Samuels, David, "In a Ruined Country: How Yasir Arafat Destroyed Palestine," *The Atlantic,* Vol. 296, No. 2, September 2005, pp. 60–91. As of August 23, 2006:
http://www.theatlantic.com/doc/prem/200509/samuels.

Shapira, Shimon, "The Origins of Hizballah," *The Jerusalem Quarterly,* Vol. 46, Spring 1988, pp. 115–130.

Shulsky, Abram N., and Gary J. Schmitt, *Silent Warfare: Understanding the World of Intelligence,* Washington, DC: Potomac Books, 2002.

Simon, Steven N., and Jonathan Stevenson, "Declawing the 'Party of God': Toward Normalizing in Lebanon," *World Policy Journal,* Vol. 18, No. 2, Summer 2001, pp. 31–42.

Tessler, Mark, *A History of the Israeli-Palestinian Conflict,* Bloomington, IN: University of Indiana Press, 1994.

Van Evera, Stephen, "Hypotheses on Nationalism and War," *International Security,* Vol. 18, No. 4, Spring 1994, pp. 5–39.

Wege, Carl Anthony, "Hizbollah Organization," *Studies in Conflict & Terrorism,* Vol. 17, No. 2, 1994, pp. 151–164.

Weiner, Myron, "Bad Neighbors, Bad Neighborhoods: An Inquiry into the Causes of Refugee Flows," *International Security,* Vol. 218, No. 1, Summer 1996, pp. 5–42.

Wolf, Eric, *Peasant Wars of the Twentieth Century,* Norman, OK: University of Oklahoma Press, 1999.